Everyday Mathematics®

The University of Chicago School Mathematics Project

Student Math Journal
Volume 1

Grade

Mc Graw Hill **Wright Group**

The McGraw-Hill Companies

The University of Chicago School Mathematics Project (UCSMP)

Max Bell, Director, UCSMP Elementary Materials Component; Director, *Everyday Mathematics* First Edition
James McBride, Director, *Everyday Mathematics* Second Edition
Andy Isaacs, Director, *Everyday Mathematics* Third Edition
Amy Dillard, Associate Director, *Everyday Mathematics* Third Edition

Authors

Max Bell	Amy Dillard	Kathleen Pitvorec
Jean Bell	Robert Hartfield	Peter Saecker
John Bretzlauf	Andy Isaacs	
Mary Ellen Dairyko*	James McBride	

*Third Edition only

Technical Art
Diana Barrie

Teachers in Residence
Lisa Bernstein, Carole Skalinder

Editorial Assistant
Jamie Montague Callister

Contributors
Carol Arkin, Robert Balfanz, Sharlean Brooks, James Flanders, David Garcia, Rita Gronbach, Deborah Arron Leslie, Curtis Lieneck, Diana Marino, Mary Moley, William D. Pattison, William Salvato, Jean Marie Sweigart, Leeann Wille

Photo Credits
©Tim Flach/Getty Images, cover; ©Fotosearch, p. vii; Getty Images, cover, *bottom left;* ©Brian Hagiwara/Brand X Pictures/Getty Images, p. iii.

www.WrightGroup.com

Send all inquiries to:
Wright Group/McGraw-Hill
P.O. Box 812960
Chicago, IL 60681

ISBN 0-07-604567-6

2 3 4 5 6 7 8 9 CPC 12 11 10 09 08 07 06

The *McGraw-Hill* Companies

Contents

UNIT 2 — Adding and Subtracting Whole Numbers

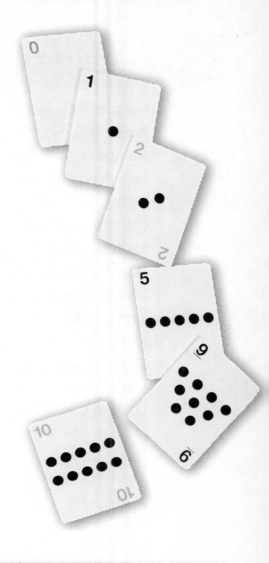

UNIT 3 — Linear Measures and Area

UNIT 4   Multiplication and Division

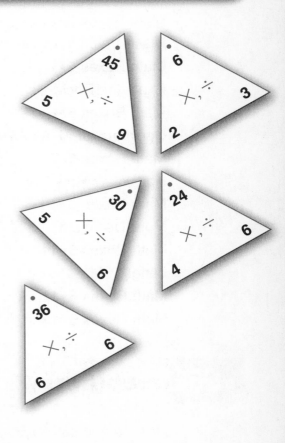

UNIT 5   Place Value in Whole Numbers and Decimals

UNIT 6   Geometry

Activity Sheets

LESSON 1·1   Number Sequences

Complete the number sequences.

Unit

1. 428, 429, _____, 431, _____, _____, _____, _____,

   436, _____, ...

2. 918, 919, _____, _____, 922, _____, _____, _____, 926, ...

3. _____; 1,416; _____; _____; 1,419; _____; _____; ...

4. _____, 311, _____, _____, 341, _____, _____, ...

5. _____; 4,326; _____; _____; 4,356; _____; ...

Try This

6. 7,628; _____; 7,828; _____; _____; 8,128

LESSON 1·1   A Numbers Hunt

Look for numbers in your classroom. Write the numbers in the table.
Look for numbers that you cannot see but you can find by counting or
measuring. Record these numbers, too.

Number	Unit (if there is one)	What does the number tell you?	How did you find the number? (count, measure, another way?)
16	Crayons	Tells how many crayons are in a box	Number is on the box
30	Inches	Height of my desk	Measured my desk

LESSON 1·2   Number-Grid Puzzles

1. Follow your teacher's directions.

541			544						550
551		553			556			559	
	562			565					570
			574			577			
581				585			588		
		593						599	
	602				606				
			614						620

Fill in the missing numbers.

2.

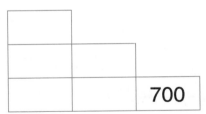

54

3.

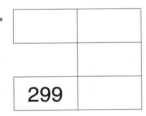

69

78

4.

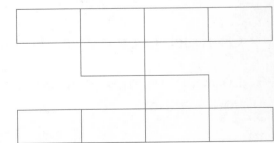

307

316 ← **317** +1, 318

327

5.

700

6.

299

7.

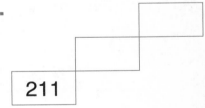

211

Do your own.

8.

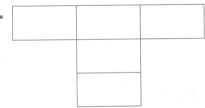

9.

three   **3**

LESSON 1·3 **Looking up Information**

1. Turn to page 246 in your *Student Reference Book*.

   How many yards are there in 1 mile? _____ yards

Work with a partner. Use your *Student Reference Book* for Questions 3–6.

2. Write your partner's first name. _____

   Write your partner's last name. _____

3. Look up the word **circumference** in the Glossary. Copy the definition.

   _____

4. Read the essay "Tally Charts."

   a. Then solve the Check Your Understanding problems.

      Problem 1: _____

      Problem 2: _____

   b. Check your answers in the Answer Key.

   c. Describe what you did to find the essay.

      _____

      _____

5. Find the Measurement section. Which of the following units
   of length is about the same length as a person's height? _____

   a. yard   b. thumb   c. fathom   d. cubit   e. hand   f. foot

   On which page did you find the answer? _____

6. Look up the rules of the game *Less Than You!* Play the game with
   your partner.

LESSON 1·4   Using Mathematical Tools

In Problems 1 and 2, record the time shown on the clocks. In Problem 3, draw the minute hand and the hour hand to show the time.

1.

_____

2.

_____

3.

6:10

Use your ruler.

4. Measure the line segment.   about _____ inches

_____

5. Draw a line segment 10 centimeters long.

Use your calculator to do these problems.

6. 23,573 + 859 + 6,051 = _____

7. 20,748 − 8,967 = _____

8. 466 × 38 = _____

9. 1,978 ÷ 23 = _____

Use your Pattern-Block Template to draw the following shapes:

10. a rhombus   **11.** a hexagon   **12.** a trapezoid

Try This

13. Which of the shapes in Problems 10–12 are quadrangles?

_____

LESSON 1·5 Displaying Data

1. How many first names are there? _____ last names? _____

2. Make a tally chart for either first names or last names.

Number of Letters	Number of Children
2	
3	
4	
5	
6	
7	
8	
9 or more	

_____ Names

3. How many letters does the longest name have? _____ letters
   The number of letters in the longest name is called the **maximum.**

4. How many letters does the shortest name have? _____ letters
   The number of letters in the shortest name is called the **minimum.**

5. What is the **range** of the numbers of letters? _____ letters
   The range is the difference between the largest and smallest numbers.

6. What is the **mode** of the set of data? _____ letters
   The number that occurs most often is called the mode.

7. What is the **median** of the set of data? _____ letters
   (*Hint:* Look in your *Student Reference Book.*)

LESSON 1·5   Displaying Data *continued*

8. Make a bar graph for your set of data.

Title: _____

LESSON 1·5 **Math Boxes**

---

1. Write the numbers that are 10 less and 10 more.

_____ 38 _____

_____ 245 _____

_____ 367 _____

_____ 1,587 _____

2. Put these numbers in order from smallest to largest.

306   _____ (smallest)

296   _____

496   _____

936   _____ (largest)

---

3. Write 5 names in the 25-box.

25

SRB
14 15

4. Draw hands on the clock to show 6:45.

---

5. Lara brought 14 candies to school. She gave away 7 during recess. How many candies does she have now?

_____ candies

6. Add.

$0 + 7 =$ _____

$5 + 1 =$ _____

$3 + 3 =$ _____

_____ $= 4 + 7$

_____ $= 9 + 6$

Unit

SRB
50 51

LESSON 1·6 Name-Collection Boxes

Work with a partner.

1. Write at least 10 names in the 20-box.

20

2. Three names do not belong in this box. Cross them out. Then write the name of the box on the tag.

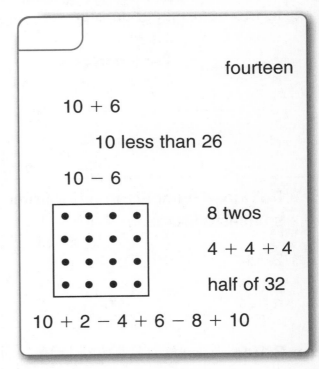

fourteen

10 + 6

10 less than 26

10 − 6

8 twos

4 + 4 + 4

half of 32

10 + 2 − 4 + 6 − 8 + 10

Do these on your own.

3. Write at least 10 names in the 24-box.

24

4. Make up your own box. Write at least 10 names.

LESSON 1·6 **Math Boxes**

---

1. Measure the line segment to the nearest inch.

_____ inches

to the nearest centimeter

_____ centimeters

_____

SRB
137 138 144

---

2. Fill in the missing numbers.

	352

SRB
7 8

---

3. Put these numbers in order from smallest to largest.

2,764   _____

8,596   _____

2,199   _____

5,096   _____

SRB
20

---

4. Count back by 2s.

104, _____, _____, _____,

96, _____, _____, _____,

_____, _____, _____, _____,

_____, _____, _____, _____

---

5. 6,347

What value does the 6 have? <u>6,000</u>

What value does the 3 have? _____

What value does the 4 have? _____

What value does the 7 have? _____

SRB
18 19

---

6. Marque had $6. His mother gave him $8. How much money does Marque have now?

$_____

LESSON 1·7   **Can You Be Sure?**

1. Make a list of things you are *sure* will happen.

   _____

   _____

   _____

   _____

   _____

   _____

2. Make a list of things you are sure will *not* happen.

   _____

   _____

   _____

   _____

   _____

3. Make a list of things you think *might* happen but you are not sure about.

   _____

   _____

   _____

   _____

   _____

Math Boxes

---

1. Write the numbers.

10 less		10 more
_____	100	_____
_____	200	_____
_____	300	_____
_____	1,000	_____

2. Put these numbers in order from smallest to largest.

4,306 _____

4,296 _____

3,496 _____

2,936 _____

SRB
20

---

3. Write at least 5 names in the 75-box.

SRB
14 15

4. What time does the clock show?

_____

---

5. Allison swam 16 laps in the pool. Carmen swam 9. How many more laps did Allison swim than Carmen? Fill in the circle next to the best answer.

○ **A.** 25 laps

○ **B.** 23 laps

○ **C.** 16 laps

○ **D.** 7 laps

6. Add.

Unit

$3 + 6 =$ _____

_____ $= 5 + 7$

$8 + 6 =$ _____

$9 + 9 =$ _____

$6 + 4 =$ _____

SRB
50 51

LESSON 1·8 **Finding Differences**

1	2	3	4	5	6	7	8	9	0
									10
11	12	13	14	15	16	17	18	19	20
21	22	23	24	25	26	27	28	29	30
31	32	33	34	35	36	37	38	39	40
41	42	43	44	45	46	47	48	49	50
51	52	53	54	55	56	57	58	59	60
61	62	63	64	65	66	67	68	69	70
71	72	73	74	75	76	77	78	79	80
81	82	83	84	85	86	87	88	89	90
91	92	93	94	95	96	97	98	99	100
101	102	103	104	105	106	107	108	109	110

Use the number grid to help you solve these problems.

1. Which is less, 83 or 73? _____ How much less? _____

2. Which is less, 13 or 34? _____ How much less? _____

3. Which is more, 90 or 55? _____ How much more? _____

4. Which is more, 44 or 52? _____ How much more? _____

Find the **difference** between each pair of numbers.

5. 71 and 92 _____

6. 26 and 46 _____

7. 30 and 62 _____

8. 48 and 84 _____

9. 43 and 60 _____

10. 88 and 110 _____

thirteen **13**

LESSON 1·8

Math Boxes

1. Measure the line segment

to the nearest inch. _____ in.

to the nearest centimeter.

_____ cm

_____

SRB
137 138
143 144

2. Fill in the missing numbers.

632

644

SRB
7 8

3. Put these numbers in order from largest to smallest.

2,764   _____

596   _____

2,199   _____

8,096   _____

SRB
20

4. Count by 2s.

1,012; 1,014; _____;

_____; _____; _____;

_____; _____; _____;

_____; _____; _____;

_____; _____; _____

5. 1,942

What value does the 4 have? _____

What value does the 9 have? _____

What value does the 1 have?

_____

What value does the 2 have? _____

SRB
18 19

6. Andre scored 7 points. Tina scored 5 points. How many points did they score altogether? Fill in the circle next to the best answer.

○ **A.** 2 points

○ **B.** 12 points

○ **C.** 17 points

○ **D.** 35 points

LESSON 1·9 Using a Calculator

Math Message

Use your calculator.

1. Sharon read the first 115 pages of her book last week. She read the rest of the book this week. If she read 86 pages this week, how many pages long is her book?

   Answer: Her book is _____ pages long.

   Number model: _____

2. The paper clip was invented in 1868. The stapler was invented in 1900. How many years after the paper clip was the stapler invented?

   Answer: The stapler was invented _____ years later.

   Number model: _____

3. $28 + 64 + 39 + 42 =$ _____   4. $2,648 - 1,576 =$ _____

Calculator Practice

Use your calculator.

5. Begin at 25. Count up by 6s. Record your counts below.

   25 ___ ___ ___ ___ ___ ___ ___ ___ ___ ___

6. Begin at 90. Count back by 9s.

   90 ___ ___ ___ ___ ___ ___ ___ ___ ___

   Solve the calculator puzzles. Remember to add or subtract to find the "Change to" number.

7.

Enter	Change to	How?
42	52	_____
61	41	_____
145	105	_____

8.

Enter	Change to	How?
178	208	_____
1,604	804	_____
722	3,722	_____

LESSON 1·9   Math Boxes

---

1. How many children
chose grapes? _____ children

How many children
chose oranges? _____ children

Children's Fruit Choices

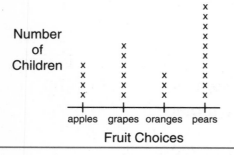

Number
of
Children

apples grapes oranges pears
Fruit Choices

SRB
77 78

---

2. Count back by 3s.

42, _____, _____, 33,

_____, _____, _____, _____,

_____, _____, _____, _____,

_____, _____, _____, _____

---

3. Use + or − to make each number
sentence true.

8 = 13 _____ 5

15 = 7 _____ 8

17 _____ 9 = 8

---

4. Circle the letter next to the coins
that do **not** show $0.89.

A Ⓠ Ⓠ Ⓠ Ⓓ Ⓟ Ⓟ Ⓟ Ⓟ

B Ⓠ Ⓠ Ⓠ Ⓓ Ⓓ Ⓟ Ⓟ Ⓟ Ⓟ Ⓟ

C Ⓠ Ⓠ Ⓓ Ⓓ Ⓓ Ⓟ Ⓟ Ⓟ
Ⓟ Ⓟ Ⓟ Ⓟ Ⓟ Ⓟ

D Ⓠ Ⓠ Ⓓ Ⓓ Ⓓ Ⓝ Ⓟ
Ⓟ Ⓟ Ⓟ

---

5. What is today's date?

_____

What will be the date in 6 days?

_____

What will be the date in 1 week?

_____

---

6. Fill in the blanks.

Unit

8 + _____ = 15

7 + _____ = 15

_____ − 8 = 7

15 − _____ = 8

SRB
50 51

LESSON 1·10 Using Coins

Math Message

1. You buy a carton of juice for 89 cents. Show two ways to pay for it with exact change. Draw Ⓟ to show pennies, Ⓝ to show nickels, Ⓓ to show dimes, and Ⓠ to show quarters.

   a. _____ b. _____

Write each of the following amounts in dollars-and-cents notation. The first one is done for you.

2. three dimes and one nickel __$0.35__

3. five dimes and seven pennies _____

4. fourteen dimes _____

5. two quarters and four pennies _____

6. three dollars and one nickel and three pennies _____

7. seven dollars and eight dimes _____

Write =, <, or >.

8. $0.68 _____ Ⓠ Ⓠ Ⓠ Ⓠ

9. Ⓓ Ⓓ Ⓝ Ⓝ Ⓝ Ⓟ Ⓟ _____ Ⓠ Ⓝ Ⓟ

10. $1.18 _____ $1.81

11. three quarters _____ three dimes

12. ten dimes _____ one dollar

13. $0.67 _____ seven dimes

> **Remember**
>
> = means *is equal to*
>
> < means *is less than*
>
> > means *is greater than*

LESSON 1·10   **Using Coins** *continued*

14. Circle the digit that represents dimes.

$ 1 7 . 6 3

15. Circle the digit that represents pennies.

$ 1 8 . 3 4

16. Circle the digit that represents dimes.

3 5 ¢

17. Jean wants to buy a carton of milk for 35¢.
How much change will she get from 2 quarters? _____

Use Ⓠ, Ⓓ, Ⓝ, and Ⓟ to show her change in two ways.

_____

Try This

Use the Drinks Vending Machine Poster on *Student Reference Book,* page 212.

18. Marcy wants to get a strawberry yogurt drink and a chocolate milk from the vending machine. She has only dollar bills.

a. If the Exact Change light is on, can she buy what she wants? _____

b. If the Exact Change light is off, how many dollar bills will she put in the

machine? _____

How much change will she get? _____

LESSON 1·10   **Math Boxes**

1. Write the number that is 10 more.

42 _____

160 _____

901 _____

59 _____

120 _____

 SRB 7

2. Ages of 9 teachers: 30, 24, 49, 50, 38, 44, 40, 35, 51

median = _____

maximum = _____

 SRB 79 80

3. Write at least 5 names in the 1-box.

1

SRB 14 15

4. Describe 2 events that are impossible.

 SRB 92

5. Put these numbers in order from smallest to largest.

7,912 _____

7,192 _____

9,271 _____

9,172 _____

SRB 20

6. Fill in the missing numbers.

_____ = 3 + 5

_____ = 8 − 5

_____ = 5 + 3

_____ = 8 − 3

Unit

 SRB 50 51

LESSON 1·11 A Shopping Trip

Use the Stationery Store Poster on *Student Reference Book,* page 214.

1. List the items you are buying in the space below. You must buy at least 3 items. You can buy 2 of the same item but list it twice.

Item	Sale Price
_____	_____
_____	_____
_____	_____

2. Estimate how many dollar bills you will need to give the shopkeeper to pay for your items.   _____ dollar bills.

3. Give the shopkeeper the dollar bills.

4. The shopkeeper calculates the total cost using a calculator.

   You owe $_____.

5. The shopkeeper calculates the change you should be getting. $_____

6. Use Ⓟ, Ⓝ, Ⓓ, Ⓠ, and $1 to show the change you got from the shopkeeper. _____

Try This

7. Henry buys one pack of batteries and a box of crayons. How much money does he save buying them on sale instead of paying the regular price?

	Regular Price	Sale Price		Difference
batteries	$_____.____	$_____.____	Regular total	$_____.____
crayons	$_____.____	$_____.____	Sale total	$_____.____
Total Cost	$_____.____	$_____.____	**Amount Saved**	$_____.____

LESSON 1·11   Coin Collections

Get your coin collection or grab a handful of coins from the classroom collection. Complete the problems below.

1. Count each kind of coin. Give a total value for each type of coin.

    _____ Ⓟ = $_____._____

    _____ Ⓝ = $_____._____

    _____ Ⓓ = $_____._____

    _____ Ⓠ = $_____._____

2. What is the total value of all the coins? You may use a calculator.

    Total value = $_____._____

3. In the space below, draw a picture of your total. Use as few $1, Ⓠ, Ⓓ, Ⓝ, and Ⓟ as possible.

Try This

4. Explain how you would enter your total amount on the calculator.

    _____

    _____

5. How much money would you need to go up to the next dollar amount? (*Hint:* A dollar amount is $1.00, $2.00, $3.00, and so on.)

    _____

    _____

6. Explain how you would go up to the next dollar amount without clearing your calculator.

    _____

    _____

Math Boxes

1. How many children
chose apples? _____ children

How many children
chose pears? _____ children

Children's Fruit Choices

```
                          x
                          x
                          x
Number                    x
of            x           x
Children      x           x
         x    x           x
         x    x    x      x
         x    x    x      x
         x    x    x      x
        ———  ———  ———    ———
       apples grapes oranges pears
```
Fruit Choices

SRB
77 78

2. Count by 10s.

23, _____, _____, 53, _____,

_____, _____, _____, _____,

_____, _____, _____, _____

3. Use + or − to make each number
sentence true.

5 _____ 6 = 11

6 _____ 5 = 1

14 = 7 _____ 7

Unit

4. Use Ⓠ, Ⓓ, Ⓝ, and Ⓟ. Show
$1.48 in two ways.

5. What is today's date?

_____

What will be the date in 11 days?

_____

What will be the date in 2 weeks?

_____

6. Complete the fact triangle.

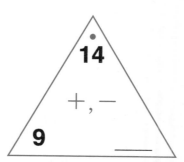

SRB
54 55

LESSON 1·12   Frames and Arrows

Math Message

Find the pattern. Fill in the missing numbers.

1. 37, 40, 43, _____, _____, _____

3. _____, 11, 15, _____, 23, _____

2. 27, 25, _____, 21, _____, _____

4. _____, _____, 36, 33, _____, 27

Frames and Arrows

5.

Rule +5¢

| 10¢ | | | 25¢ | |

6.

Rule +10

| 32 | | 52 | | | 82 |

7.

Rule

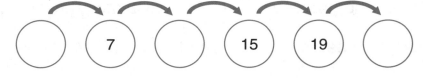

| | 7 | | 15 | 19 | |

8.

Rule Double

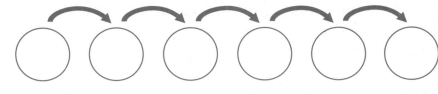

| 2 | | 8 | | | 64 |

9. Make up one of your own.

Rule

LESSON 1·12 Patterns

Complete the number-grid puzzles.

1.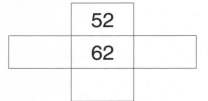

2.

3.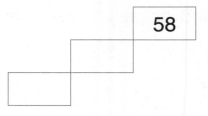

---

4. Draw dots to show what comes next.

                • • •

       • •   • • •

  •   • •   • • •

5. Janie owns a magic calculator. When someone enters a number and then presses the ⊜ key, it changes the number. Here is what happened:

   ♦ Tom entered 15. He pressed ⊜ and the calculator showed 5.

   ♦ Mary entered 12. She pressed ⊜ and the calculator showed 2.

   ♦ Regina entered 27. She pressed ⊜ and the calculator showed 17.

6. What do you think the calculator will show if Janie enters 109 and ⊜? _____

7. Explain how you know. _____

Try This

8. The numbers below have a pattern. Fill in the missing numbers.
   Be careful: The same thing does not always happen each time.

   4, 14, 24, 22, 32, 42, 40, 50, 60, 58, _____, _____, _____

9. Describe the pattern. _____

_____

_____

LESSON 1·12 | **Math Boxes**

---

1. Write the number that is 100 more.

237 _____

614 _____

994 _____

2,462 _____

3,965 _____

SRB
7

2. Median number of books read: ____ books

Maximum number of books read: ____ books

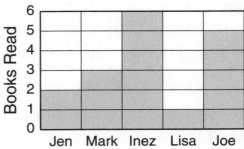

SRB
79 80

---

3. Write 5 names in the 0-box.

| **0** |
| |

SRB
14 15

4. Describe 2 events that are almost sure to happen.

_____

_____

_____

_____

_____

---

5. Which group is in order from largest to smallest? Fill in the circle next to the best answer.

Ⓐ 4,039; 4,040; 4,409; 4,009

Ⓑ 4,409; 4,040; 4,039; 4,009

Ⓒ 4,009; 4,039; 4,040; 4,409

Ⓓ 4,040; 4,039; 4,009; 4,409

6. Fill in the blanks.

Unit
[   ]

4 + 2 = _____

6 − _____ = 2

_____ + 4 = 6

6 − 2 = _____

SRB
50 51

LESSON
1·13   **Finding Elapsed Times**

Use your toolkit clock to help you solve these problems.

1. Ava leaves to go swimming at 4:05 and
   returns at 5:25. How long has she been
   gone?

   _____

2. Deven rides his bike 37 miles. He rides
   from 10:15 A.M. until 3:50 P.M. How long
   does it take him to ride 37 miles?

   _____

3. LaToya leaves for school at the time
   shown on the first clock. She returns
   home at the time shown on the second
   clock. How long is LaToya away
   from home?

   _____

Try This

4. Gregory baked cookies for a class party.
   He baked several different kinds. He
   began baking at the time shown on the
   first clock and finished at the time shown
   on the second clock. How long did it
   take Gregory to bake the cookies?

   _____

Explain how you figured out the answer.

_____

_____

_____

_____

LESSON 1·13 Sunrise and Sunset Record

Date	Time of Sunrise	Time of Sunset	Length of Day
			hr min
			hr min
			hr min
			hr min
			hr min
			hr min
			hr min
			hr min
			hr min
			hr min
			hr min
			hr min
			hr min
			hr min
			hr min
			hr min
			hr min
			hr min
			hr min
			hr min
			hr min
			hr min
			hr min
			hr min

LESSON 1·13   Math Boxes

1. How many children like lions?_____

How many children like tigers?_____

Animal choice	Number of Children
Bears	////
Lions	~~HHH~~ /
Crows	///
Tigers	~~HHH~~ ~~HHH~~

2. Count back by 4s.

104, _____, _____, _____, 88

_____, _____, _____, _____,

_____, _____, _____, _____,

_____, _____, _____

3. Use + or − to make each number sentence true.

9 = 3 _____ 3 _____ 3

12 = 4 _____ 4 _____ 4

18 = 9 _____ 9

Unit

4. Draw the bills and coins to show $2.43 in two ways.

5.

OCTOBER						
Su	M	Tu	W	Th	F	Sa
		1	2	3	4	5
6	7	8	9	10	11	12

If October 6 is on a Sunday, what are the dates for the *next* two Sundays?

_____

_____

6. Complete the Fact Triangle.

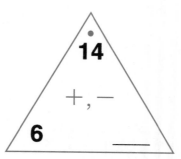
14

+, −

6   _____

SRB
54 55

LESSON 1·14   Math Boxes

---

1. Complete the fact family.

$4 + 5 =$ _____

$5 +$ _____ $= 9$

$9 - 4 =$ _____

_____ $- 5 =$ _____

Unit
[ ]

SRB
50 51

---

2. Ava scored 9 goals this season. Jamar scored 6 goals. How many goals did they score altogether?

_____ goals

---

3. Fill in the blanks.

$7 -$ _____ $= 7$

$13 +$ _____ $= 13$

$9 + 1 =$ _____

$8 + 1 =$ _____

Unit
[ ]

SRB
50 51

---

4. Nico walks 6 blocks to school. Cyrus walks 4 blocks to school. How many blocks do they walk in all?

_____ blocks

---

5. Fill in the blanks.

_____ $= 9 - 8$

_____ $= 7 - 6$

$1 = 5 -$ _____

$1 =$ _____ $- 7$

Unit
[ ]

SRB
50 51

---

6. Jeri brought 10 erasers to school. She gave 6 erasers to her friends. How many erasers does she have left?

_____ erasers

Fact Families and Number Families

Complete the Fact Triangles. Write the fact families.

1.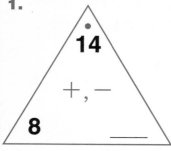

2.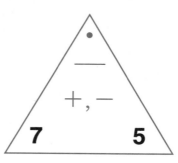

3.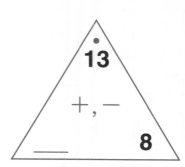

_____ + _____ = _____   _____ = _____ + _____   _____ + _____ = _____

_____ + _____ = _____   _____ = _____ + _____   _____ + _____ = _____

_____ − _____ = _____   _____ = _____ − _____   _____ − _____ = _____

_____ − _____ = _____   _____ = _____ − _____   _____ − _____ = _____

Complete the number triangles. Write the number families.

4.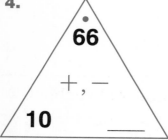

5.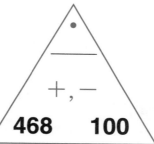

6.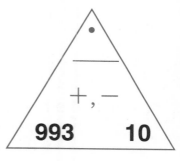

_____ = _____ + _____   _____ = _____ + _____   _____ + _____ = _____

_____ = _____ + _____   _____ = _____ + _____   _____ + _____ = _____

_____ = _____ − _____   _____ = _____ − _____   _____ − _____ = _____

_____ = _____ − _____   _____ = _____ − _____   _____ − _____ = _____

LESSON 2·1 Math Boxes

1. Put these numbers in order from smallest to largest.

1,532   _____

1,253   _____

1,325   _____

5,321   _____

SRB 20

2. Fill in the missing numbers.

		975	

SRB 7 8

3. Write the numbers that are 10 less and 10 more than each given number.

	10 less	10 more
368	_____	_____
789	_____	_____
1,999	_____	_____
40,870	_____	_____

SRB 7, 18, 19

4. I spent $3.25 at the store. I gave the cashier a $5.00 bill. How much change should I have received?

_____

5. About what time is it? Fill in the circle next to the best answer.

Ⓐ 1:35

Ⓑ 7:08

Ⓒ 1:40

Ⓓ 2:35

6. Measure the line segment in inches.

_____ inches

_____

SRB 143 144

LESSON 2·2 Using Basic Facts to Solve Fact Extensions

Fill in the unit box. Complete the fact extensions.

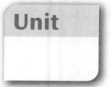

Unit

1. _____ = 12 − 7

   _____ = 120 − 70

   _____ = 1,200 − 700

2. 8 + 3 = _____

   80 + 30 = _____

   800 + 300 = _____

3. _____ = 7 + 6

   _____ = 70 + 60

   _____ = 700 + 600

Complete the fact extensions.

4. _____ = 6 + 8

   _____ = 16 + 8

   _____ = 56 + 8

5. 14 − 9 = _____

   24 − 9 = _____

   54 − 9 = _____

6. _____ = 17 − 11

   _____ = 27 − 11

   _____ = 47 − 11

Use addition or subtraction to complete these problems on your calculator. You may also use a number grid, base-10 blocks, or counters.

7. Enter	Change to	How?
33	40	_____
80	73	_____
80	23	_____

8. Enter	Change to	How?
430	500	_____
700	640	_____
1,000	400	_____

9. Why is it important to know the basic addition and subtraction facts?

   _____

   _____

LESSON 2·2

Math Boxes

1. Complete the Fact Triangle. Write the fact family.

_____ + _____ = _____

_____ + _____ = _____

_____ − _____ = _____

_____ − _____ = _____

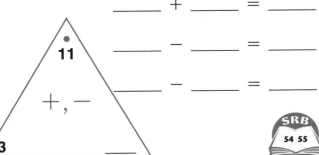

11

+, −

3

SRB
54 55

2. Choose the best answer.

The school chorus has 28 second graders and 34 third graders. How many children are in the chorus?

◯ 52 children ◯ 14 children

◯ 12 children ◯ 62 children

3. Use your calculator. Write the answers in dollars and cents.

$0.85 + 53¢ = _____

$2.08 + $5.01 = _____

64¢ + $1.73 = _____

37¢ + 26¢ = _____

4. Find the rule. Fill in the empty frames.

Rule

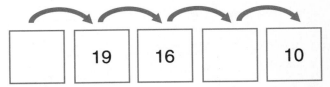

| | 19 | 16 | | 10 |

SRB
200 201

5. The mode for the number of

books read is _____.

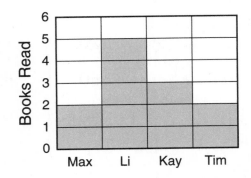

SRB
81 82

6. Write these units of measure in order from smallest to largest.

mile _____ ← smallest

foot _____

yard _____

inch _____ ← largest

SRB
148

LESSON 2·3 "What's My Rule?"

Fill in the blanks.

1.

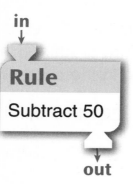

Rule
Subtract 50

in	out
100	
50	
70	
150	
200	

2.

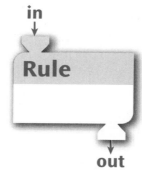

Rule

in	out
14	23
34	43
44	53
64	73
94	103

3.

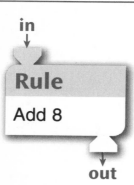

Rule
Add 8

in	out
	13
	23
	43
	73
	93

4.

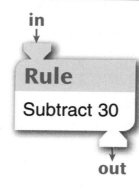

Rule
Subtract 30

in	out
	30
	50
	100
	200
	0

5.

Rule

in	out
6	13
9	
5	
4	11
	18

6.

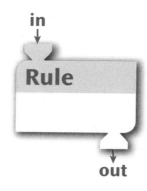

Rule

in	out
35	20
	45
20	
50	35
46	

LESSON 2·3 Math Boxes

1. Put these numbers in order from smallest to largest.

32,764 _____

8,596 _____

32,199 _____

85,096 _____

SRB 20

2. Fill in the missing numbers.

	1,073		
		1,104	

SRB 7 8

3. Write the numbers that are 100 more and 100 less than each given number.

          100 more   100 less

614 _____ _____

994 _____ _____

2,462 _____ _____

3,965 _____ _____

SRB 18 19

4. You spent $7.88 at the store. You gave the cashier a $10 bill. How much change should you receive?

_____

5. What time is it?

_____

What time will it be in 20 minutes?

_____

How many minutes until 5:15?

_____

6. Measure the line segment to the nearest $\frac{1}{2}$ inch.

_____ inches

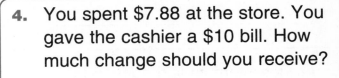

SRB 143–145

**LESSON
2·4**   # Number Stories: Animal Clutches

For each number story, write *?* for the number you want to find. Write the numbers you know in the parts-and-total diagram. Solve the problem, and write a number model.

1. Two pythons laid clutches of eggs. One clutch had 36 eggs. The other had 23 eggs. How many eggs were there in all?

   Answer the question: _____
   (unit)

   Number model: _____

   Check: How do you know your answer makes sense?

   _____

   _____

Total	
Part	**Part**

2. A queen termite laid about 6,000 eggs on Monday and about 7,000 eggs on Tuesday. About how many eggs did she lay in all?

   Answer the question: _____
   (unit)

   Number model: _____

   Check: How do you know your answer makes sense?

   _____

   _____

Total	
Part	**Part**

3. Two clutches of Mississippi alligator eggs were found. Each clutch had 47 eggs. What was the total number of eggs found?

   Answer the question: _____
   (unit)

   Number model: _____

   Check: How do you know your answer makes sense?

   _____

   _____

Total	
Part	**Part**

LESSON 2·4 Number Stories: Animal Clutches *continued*

Try This

4. An alligator clutch had 60 eggs. Only 12 hatched. How many eggs did not hatch?

Answer the question: _____
(unit)

Number model: _____

Check: How do you know your answer makes sense?

_____

_____

Total	
Part	Part

5. Scientists say a green turtle can lay about 1,800 eggs in a lifetime, but only about 400 eggs hatch overall. About how many eggs do not hatch?

Answer the question: _____
(unit)

Number model: _____

Check: How do you know your answer makes sense?

_____

_____

Total	
Part	Part

Math Boxes

1. Complete the Fact Triangle. Write the fact family.

_____ + _____ = _____

_____ + _____ = _____

_____ − _____ = _____

_____ − _____ = _____

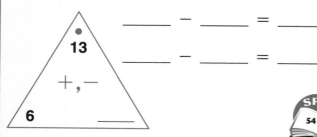

13

+, −

6 _____

SRB
54 55

2. Jonah had $52. He bought a CD for $14. How much money does he have now?

_____

Number model:

_____

3. Use your calculator. Write the answers in dollars and cents.

73¢ + $2.65 = _____

$0.65 + 47¢ = _____

$1.06 + $5.21 = _____

46¢ + 35¢ = _____

4. Fill in the empty frames.

Rule

+100

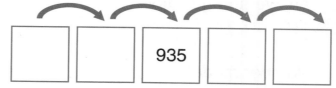

935

SRB
200 201

5. Weekly allowances:
$15, $12, $5, $8

The maximum weekly allowance is

_____ .

The minimum weekly allowance is

_____ .

The range of weekly allowances is

_____ .
SRB
79

6. Write these metric units of measure in order from smallest to largest.

centimeter _____ ← smallest

kilometer _____

millimeter _____

meter _____ ← largest

SRB
141

LESSON 2·5 Number Stories: Change-to-More and Change-to-Less

For each number story, write ? in the diagram for the number you
want to find. Then write the numbers you know in the change diagram
also. Next, solve the problem. Write the answer and a number model.

Unit

dollars

1. Ahmed had $22 in his bank account. For his
    birthday, his grandmother deposited $25 for him.
    How much money is in his bank account now?

    Answer the question: _____

    Number model: _____

    Check: How do you know your answer makes sense?

    _____

Change

| Start | | End |

_____

2. Omar had $53 in his piggy bank. He used $16
    to take his sister to the movies and buy treats.
    How much money is left in his piggy bank?

    Answer the question: _____

    Number model: _____

    Check: How do you know your answer makes sense?

    _____

Change

| Start | | End |

_____

3. Cleo had $37 in her purse. Then Jillian returned
    $9 that she borrowed. How much money does
    Cleo have now?

    Answer the question: _____

    Number model: _____

    Check: How do you know your answer makes sense?

Change

| Start | | End |

_____

Number Stories *continued*

4. Audrey had $61 in her bank account. She withdrew $48 to take on vacation. How much is left in her account?

Answer the question: _____

Number model: _____

Check: How do you know your answer makes sense?

_____

Try This

5. Trung had $15 in his piggy bank. After his birthday, he had $60 in his bank. How much money did Trung get as birthday presents?

Answer the question: _____

Number model: _____

Check: How do you know your answer makes sense?

_____

6. Nikhil had $40 in his wallet when he went to the carnival. When he got home, he had $18. How much did he spend at the carnival?

Answer the question: _____

Number model: _____

Check: How do you know your answer makes sense?

_____

_____

LESSON 2·5 **Math Boxes**

---

1. Use addition or subtraction to complete these problems on your calculator.

Enter	Change to	How?
366	66	_____
894	2,894	_____
3,775	3,175	_____
27,581	28,581	_____

SRB 18, 19, 264

---

2. "What's My Rule?"

in	out
10	
21	
32	
	60

in ↓

Rule

Add 4

↓ out

SRB 203 204

---

3. 2,345

the 2 means _____

the 3 means _____

the 4 means _____

the 5 means _____

SRB 18 19

---

4. Write 5 names in the 120-box.

120

SRB 14 15

---

5. Lily had 33 rings in one box and 29 in another. How many did she have in all?

_____ rings

Total	
Part	**Part**

SRB 256 257

---

6. How many squares are shaded? Fill in the oval for the best answer.

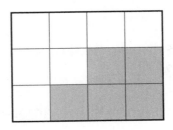

⬭ 12   ⬭ 7   ⬭ 9   ⬭ 5

LESSON 2·6   Temperature Differences

Use the map on page 220 in the *Student Reference Book* to answer
Problems 1–4. Write ? on the diagram for the number you want to find.
Write the numbers you know in the comparison diagram. Then solve
the problem. Write the answer and a number model.

1. What is the difference between the normal high
   and low temperatures for San Francisco?

   Answer the question: _____°F

   Number model:

   _____

   Check: How do you know your answer makes sense?

   _____

   _____

Quantity

Quantity

   Difference

2. What is the difference between the normal high
   and low temperatures for Seattle?

   Answer the question: _____°F

   Number model:

   _____

   Check: How do you know your answer makes sense?

   _____

   _____

Quantity

Quantity

   Difference

3. Which city has the *largest* difference between the normal high and low
   temperatures?

   _____ What is the difference? _____°F

4. Which city has the *smallest* difference between the normal high and low
   temperatures?

   _____ What is the difference? _____°F

LESSON 2·6

National High/Low Temperatures Project

Date	Highest Temperature (maximum)		Lowest Temperature (minimum)		Difference (range)
	Place	Temperature	Place	Temperature	
		°F		°F	°F
		°F		°F	°F
		°F		°F	°F
		°F		°F	°F
		°F		°F	°F
		°F		°F	°F
		°F		°F	°F
		°F		°F	°F
		°F		°F	°F
		°F		°F	°F
		°F		°F	°F
		°F		°F	°F
		°F		°F	°F
		°F		°F	°F
		°F		°F	°F
		°F		°F	°F
		°F		°F	°F
		°F		°F	°F
		°F		°F	°F
		°F		°F	°F

LESSON 2·6 **Math Boxes**

---

1. Complete the fact extensions.

$13 = 6 + 7$

_____ $= 16 + 7$

_____ $= 26 + 7$

_____ $= 106 + 7$

_____ $= 136 + 7$

Unit
```
┌──────────┐
│  Unit    │
│          │
│          │
└──────────┘
```

2. Fill in the blanks.

_____ $+ 9 = 50$

$24 + $ _____ $= 30$

_____ $= 70 - 8$

_____ $+ 73 = 80$

```
┌──────────┐
│  Unit    │
│          │
│          │
└──────────┘
```

---

3. About what time is it? Fill in the circle next to the best answer.

○ **A.** 5:30   ○ **B.** 5:05

○ **C.** 6:25   ○ **D.** 5:25

4. Find the rule and complete the table.

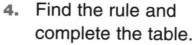

in	out
117	112
119	
	116
	131

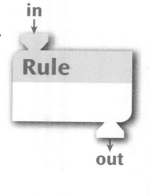

Rule

---

5. A vendor sells about 800 ice-cream bars every day. About how many ice-cream bars does the vendor sell in 2 days?

_____ ice-cream bars

Total	
Part	**Part**

6. Measure the line segment to the nearest centimeter.

_____ cm

_____

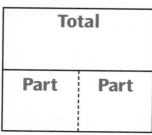

LESSON 2·7   Addition Methods

Make a ballpark estimate. Write a number model to show your estimate. Choose at least two problems to solve using the partial-sums method and show your work. You may choose any method you wish to solve the other problems.

Unit

miles

Example:

Ballpark estimate:

$$300+400 = 700$$

Partial-Sums Method

	100s	10s	1s
	3	2	9
+	4	1	8
	7	0	0
		3	0
+		1	7
	7	4	7

---

1. Ballpark estimate:

_____

```
   43
+  26
```

---

2. Ballpark estimate:

_____

```
   90
+  37
```

---

3. Ballpark estimate:

_____

```
  378
+ 401
```

---

4. Ballpark estimate:

_____

```
  172
+ 109
```

---

5. Ballpark estimate:

_____

```
   87
+ 113
```

LESSON 2·7 Math Boxes

1. Use addition or subtraction to complete these problems on your calculator.

Enter	Change to	How?
173	873	_____
4,501	1,501	_____
5,604	6,604	_____
9,646	9,346	_____

SRB 18, 19, 264

2. "What's My Rule?"

in	out
4	
	12
0	
	21

in ↓

Rule

out

SRB 203 204

3. In 1,532,

the 1 means _____

the 5 means _____

the 3 means _____

the 2 means _____

SRB 18 19

4. Write at least 5 names for 1,000.

1,000

SRB 14 15

5. Austin read his book for 45 minutes on Monday and for 25 minutes on Tuesday. How many more minutes did he read on Monday?

Quantity	

_____ minutes

Quantity	

_____
Difference

SRB 258

6. How long is the fence around the flowers?

_____ feet

3 feet

2 feet   2 feet

3 feet

SRB 150 151

LESSON 2·8 Subtraction Methods

Make a ballpark estimate. Write a number model to show your estimate. Choose at least two problems to solve using the counting-up method and show your work. You may choose any method you wish to solve the other problems.

Unit

lunches

Example:
Ballpark estimate:

$$230 - 200 = 30$$

$$\begin{array}{r} 234 \\ -\ 187 \\ \hline \end{array}$$

Counting-Up Method

$$\begin{array}{r} 187 \\ +\ ③ \\ \hline 190 \\ +\ ⑩ \\ \hline 200 \\ +\ ㉚ \\ \hline 230 \\ +\ ④ \\ \hline 234 \end{array}$$

$$3 + 10 + 30 + 4 = 47$$

---

1. Ballpark estimate:

_____

$$\begin{array}{r} 63 \\ -\ 37 \\ \hline \end{array}$$

---

2. Ballpark estimate:

_____

$$\begin{array}{r} 91 \\ -\ 46 \\ \hline \end{array}$$

---

3. Ballpark estimate:

_____

$$\begin{array}{r} 129 \\ -\ 112 \\ \hline \end{array}$$

---

4. Ballpark estimate:

_____

$$\begin{array}{r} 283 \\ -\ 256 \\ \hline \end{array}$$

---

5. Ballpark estimate:

_____

$$\begin{array}{r} 752 \\ -\ 387 \\ \hline \end{array}$$

LESSON 2·8   Name-Collection Boxes

1. Three names do not belong. Mark them with a big **X**.

100	$1,680 - 1,580$
$25 + 25 + 25$	80
$30 + 70$	$\begin{array}{r} +30 \\ \hline \end{array}$
$\begin{array}{r} 63 \\ +37 \\ \hline \end{array}$	$\begin{array}{r} 1,000 \\ -100 \\ \hline \end{array}$
2 fifties	$\begin{array}{r} 9,999 \\ -9,899 \\ \hline \end{array}$
$48 + 52$	

2. Write at least 10 names for 40.

40

3. Write at least 10 names for 200.

200

4. Write at least 10 names for 1,000.

1,000

LESSON 2·8   **Math Boxes**

---

1. Complete the fact extensions.

Unit _____

$6 + 5 = 11$

$16 + 5 =$ _____

$26 + 5 =$ _____

$86 + 5 =$ _____

$126 + 5 =$ _____

2. Fill in the blanks.

Unit

days

_____ $+ 53 = 60$

$90 = 3 +$ _____

$132 = 140 -$ _____

$198 +$ _____ $= 210$

---

3. What time does the clock show?

_____

What time will it be in 30 minutes?

_____

4. Find the rule and complete the table.

in	out
102	122
130	
	184
	193
188	

in ↓

Rule _____

out

SRB 203 204

---

5. Corey had $75. He bought a new baseball for $18. How much money does he have now?

_____

Number model:

_____

Start	Change →	End

6. Measure the line segment to the nearest $\frac{1}{2}$ centimeter.

_____ cm

_____

SRB 137–139

LESSON 2·8 Subtraction Methods

Make a ballpark estimate. Write a number model to show your estimate. Choose at least two problems to solve using the trade-first method and show your work. You may choose any method you wish to solve the other problems.

Unit

dollars

Example:

Ballpark estimate:

$$250 - 200 = 50$$

Trade-First Method

100s	10s	1s
1	14	
2̸	4̸	7
−1	8	6
	6	1

1. Ballpark estimate:

_____

$$
\begin{array}{r}
74 \\
-\ 29 \\
\hline
\end{array}
$$

2. Ballpark estimate:

_____

$$
\begin{array}{r}
96 \\
-\ 37 \\
\hline
\end{array}
$$

3. Ballpark estimate:

_____

$$
\begin{array}{r}
208 \\
-\ 106 \\
\hline
\end{array}
$$

4. Ballpark estimate:

_____

$$
\begin{array}{r}
271 \\
-\ 248 \\
\hline
\end{array}
$$

5. Ballpark estimate:

_____

$$
\begin{array}{r}
826 \\
-\ 172 \\
\hline
\end{array}
$$

LESSON 2·9   Number Stories with Several Addends

1. José bought milk for 35 cents, apple juice for 55 cents, grape juice for 45 cents, and orange juice for 65 cents.
How much money did he spend?

Total			
Part	Part	Part	Part

Answer the question: _____
(unit)

Number model:

_____

Check: How do you know your answer makes sense?

_____

_____

_____

2. Michelle drove from Houston, Texas to Wichita, Kansas. On the first day, she drove 245 miles. On the second day, she drove 207 miles. On the third day, she drove 158 miles and arrived in Wichita.
How many miles did she drive in all?

Total		
Part	Part	Part

Answer the question: _____
(unit)

Number model:

_____

Check: How do you know your answer makes sense?

_____

_____

LESSON 2·9 Number Stories with Several Addends *continued*

3. Zookeepers watched a clutch of 54 python eggs. On the first day, 18 eggs hatched. On the next day, 11 more hatched. How many eggs did not hatch?

Total		
Part	**Part**	**Part**

Answer the question: _____
(unit)

Number model:

_____

Check: How do you know your answer makes sense?

_____

_____

_____

4. Carl has $2.50 for juice or milk at lunch. On each of 2 days, he buys grape juice for 45 cents. On the third day, he buys milk for 40 cents. How much money does he have left?

Total			
Part	**Part**	**Part**	**Part**

Answer the question: _____
(unit)

Number model:

_____

Check: How do you know your answer makes sense?

_____

_____

_____

LESSON 2·9   **Math Boxes**

1. Use addition or subtraction to solve these problems on your calculator.

Enter	Change to	How?
409	909	_____
3,291	291	_____
10,538	10,138	_____
12,201	15,201	_____

SRB 18, 19, 264

2. "What's My Rule?"

in	out
14	
24	
39	
	42
	65

in ↓

Rule

Subtract 7

out

SRB 203 204

3. In 5,639, the 5 means _____.
Fill in the circle next to the best answer.

Ⓐ 500

Ⓑ 5,000

Ⓒ 50

Ⓓ 5

SRB 18 19

4. Fill in the tag. Write at least 5 names for that number.

SRB 14 15

5. Jack answered 29 questions. José answered 37 questions. How many fewer questions did Jack answer than José?

Quantity

Quantity

_____ questions

Difference

SRB 258

6. Which tool would you use to measure the following?

| yardstick | ruler | thermometer |

Temperature _____

Height of the ceiling _____

Length of your thumb _____

LESSON 2·10   **Math Boxes**

1. Which tool would you use to measure the following items:

| meterstick | 6 in. ruler | thermometer |

Outdoor
temperature _____

Length of
your calculator _____

Height of the door _____

2. Circle the best unit of measurement.

Distance to the Galápagos Islands
kilometers   centimeters   meters

Width of your thumbnail
kilometers   centimeters   meters

Length of your *Student Reference Book*
kilometers   centimeters   meters

3. Measure the line segment to the nearest $\frac{1}{2}$ inch.

_____ in.

_____

143 144

4. Measure the line segment to the nearest $\frac{1}{2}$ centimeter.

_____ cm

_____

137 138

5. How many squares are shaded?

_____ squares

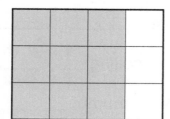

6. How long is the fence around the house?

_____ meters

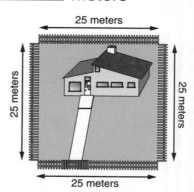

25 meters
25 meters
25 meters
25 meters

150

LESSON 3·1   Estimating and Measuring Lengths

Work with a partner. Estimate the lengths of things in the classroom in "class shoe" units. Write the estimate. Then use the class shoe strip to measure the object. Write the measurement.

1.

Object	Estimate	Measurement
	about _____ class shoes	about _____ class shoes
	about _____ class shoes	about _____ class shoes
	about _____ class shoes	about _____ class shoes
	about _____ class shoes	about _____ class shoes
	about _____ class shoes	about _____ class shoes
	about _____ class shoes	about _____ class shoes
	about _____ class shoes	about _____ class shoes
	about _____ class shoes	about _____ class shoes

2. Why is it important to use the same unit everyone else is using to measure things?

_____

_____

_____

_____

LESSON 3·1 Addition and Subtraction Practice

Make a ballpark estimate to use to check your answer. Write a number model for your estimate. Add or subtract.

Unit
pumpkin
seeds

1. Ballpark estimate:

_____

$$\begin{array}{r} 681 \\ +\ 253 \\ \hline \end{array}$$

2. Ballpark estimate:

_____

$$\begin{array}{r} 749 \\ +\ 161 \\ \hline \end{array}$$

3. Ballpark estimate:

_____

$$\begin{array}{r} 417 \\ +\ 386 \\ \hline \end{array}$$

4. Ballpark estimate:

_____

$$\begin{array}{r} 472 \\ -\ 253 \\ \hline \end{array}$$

5. Ballpark estimate:

_____

$$\begin{array}{r} 728 \\ -\ 173 \\ \hline \end{array}$$

6. Ballpark estimate:

_____

$$\begin{array}{r} 550 \\ -\ 364 \\ \hline \end{array}$$

LESSON 3·1 **Math Boxes**

---

1. Add.

 Unit

$9 + 22 + 11 =$ _____

$13 + 17 + 16 =$ _____

$24 + 6 + 9 =$ _____

---

2. Use addition or subtraction to complete these problems on your calculator.

Enter	Change to	How?
141	191	_____
406	906	_____
1,873	1,273	_____
1,462	5,462	_____

18 19

---

3. Order these numbers from smallest to largest.

1,060 _____

1,600 _____

1,006 _____

6,001 _____

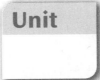
20

---

4. $148 + 45 =$ _____

Unit

Make a ballpark estimate. Write your number model.

_____ + _____ = _____

191 192

---

5. Solve using the counting-up method. Show your work.

Unit

```
   42
 − 14
 ────
```

63

---

6. Circle the event that is *unlikely* to happen.

If you toss a coin 20 times, it will always land on HEADS.

If you toss a coin 20 times, it will land on HEADS some of the times.

LESSON 3·2 Measuring Line Segments

1. Use Ruler A to measure to the nearest inch (in.).

Use Ruler D to measure to the nearest centimeter (cm).

_____

_____

2. Use Ruler B to measure to the nearest $\frac{1}{2}$ inch (in.).

Use Ruler D to measure to the nearest $\frac{1}{2}$ centimeter (cm).

_____

_____

Try This

3. Use Ruler C to measure to the nearest $\frac{1}{4}$ inch (in.).

Use Ruler D to measure to the nearest millimeter (mm).

_____

_____

_____

	Ruler A	Ruler D
	about ____ in.	about ____ cm
	about ____ in.	about ____ cm

	Ruler B	Ruler D
	about ____ in.	about ____ cm
	about ____ in.	about ____ cm
	about ____ in.	about ____ cm

	Ruler C	Ruler D
	about ____ in.	about ____ mm
	about ____ in.	about ____ mm
	about ____ in.	about ____ mm

Math Boxes

---

1. Count by 6s.

____57____ , _____ , _____ , _____ ,

____81____ , _____ , _____ , _____ ,

_____ , _____ , _____ , _____ ,

_____ , _____ , _____ , _____

---

2. Measure to the nearest $\frac{1}{4}$ inch.

about _____ in.

_____

Draw a line segment $1\frac{1}{2}$ inches long.

143–144

---

3. Write $<$, $>$, or $=$.

69 _____ 96

101 _____ 110

2Ⓠ _____ 5Ⓓ

1,000 _____ 999

13

---

4. Pamela had \$38. She spent _____ on shoes. She has \$15 left.

Change

Start		End
	_____	

254 255

---

5.   **Book Club Totals**

Number of Children

```
                        X
                  X     X
      X           X     X
      X     X     X     X
      X     X     X     X
      X     X     X     X
    |--+--+--+--+--|
    0  1  2  3  4
      Books Read
```

Maximum number of books read:

_____

77–79

---

6. Courtney has 8 pennies. She shares them equally with Nicholas. How many pennies do they each get? Fill in the circle for the best answer.

◯ A.   16 pennies

◯ B.   8 pennies

◯ C.   4 pennies

◯ D.   2 pennies

73

Measures Hunt

Find out about how long some objects are.

These objects will be **personal references.**

Use your personal references to estimate the lengths of other things.

1. Find things that are about 1 inch long, 1 foot long, and 1 yard long.
   Use a ruler, tape measure, or yardstick.
   List your objects below.

About 1 inch (in.)	**About 1 foot (ft)**	**About 1 yard (yd)**
_____	_____	_____
_____	_____	_____
_____	_____	_____
_____	_____	_____
_____	_____	_____

2. Find things that are about 1 centimeter long, 1 decimeter long, and 1 meter long.
   Use a ruler, tape measure, or meterstick.
   List your objects below.

About 1 centimeter (cm)	**About 1 decimeter (dm)**	**About 1 meter (m)**
_____	_____	_____
_____	_____	_____
_____	_____	_____
_____	_____	_____

LESSON 3·3 **Estimating Lengths**

1. Follow these steps using **U.S. customary** units: inches (in.), feet (ft), or yards (yd). Then follow these steps using **metric** units: millimeters (mm), centimeters (cm), decimeters (dm), or meters (m).

   ◆ Use personal references to estimate the measures.

   ◆ Record your estimates. Be sure to write the units.

   ◆ Measure with a ruler or tape measure. Record your measurements.

Objects	U.S. Customary Units		Metric Units	
	Estimate	Measurement	Estimate	Measurement
height of your desk				
long side of your calculator				
short side of the classroom				
distance around your head				

2. Choose your own objects to estimate and measure.

Objects	U.S. Customary Units		Metric Units	
	Estimate	Measurement	Estimate	Measurement

LESSON 3·3 Math Boxes

1. 8 + 6 = _____

   8 + 6 + 7 = _____

   8 + 6 + 7 + 5 = _____

```
  17        17         17
+  8         8          8
          +  5          5
                     + 19
```

Unit ☐

2. Use addition or subtraction to complete these problems on your calculator.

Enter	Change to	How?
267	307	_____
1,039	539	_____
1,374	1,874	_____
15,866	11,866	_____

SRB 18 19

3. Order these numbers from largest to smallest.

   1,164 _____

   1,104 _____

   1,146 _____

   1,416 _____

SRB 20

4. Estimate.
Is 82 − 49 closer to 40 or to 30?

_____

Show the number model you used.

_____ − _____ = _____

Unit ☐

SRB 192

5. Solve using the trade-first method. Show your work.

```
  66
- 38
```

Unit ☐

SRB 60 61

6. Fill in the circle for the best answer.

If you toss a quarter 100 times, you can be certain it will

Ⓐ always land on HEADS.

Ⓑ always land on either HEADS or TAILS.

Ⓒ land on HEADS 99 times.

Ⓓ land on HEADS exactly $\frac{1}{2}$ of the time.

LESSON 3·4   Perimeters of Polygons

1. Record the **perimeter** (the distance around) of your straw rectangle and parallelogram.

   rectangle: about _____ inches   parallelogram: about _____ inches

2. Use a tape measure to find each side and the perimeter.

Polygon	Each Side	Perimeter
triangle	about _____ in., about _____ in., about _____ in.	about _____ in.
triangle	about _____ in., about _____ in., about _____ in.	about _____ in.
square	about _____ in.	about _____ in.
rhombus	about _____ in.	about _____ in.
trapezoid	about _____ in., about _____ in. about _____ in., about _____ in.	about _____ in.

3. Find the perimeter, in inches, of the figures below.

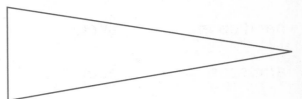

_____   _____

Try This

4. Draw each shape on the centimeter grid.
   square with perimeter = 16 cm   rectangle with perimeter = 20 cm

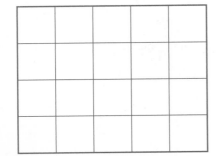

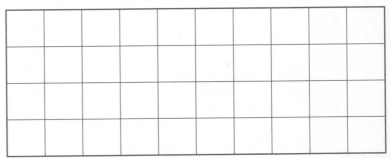

LESSON 3·4   Body Measures

Work with a partner to find each measurement to the nearest $\frac{1}{2}$ inch.

	Adult at Home	Me (Now)	Me (Later)
Date	_____	_____	_____
height	about ____ in.	about ____ in.	about ____ in.
shoe length	about ____ in.	about ____ in.	about ____ in.
around neck	about ____ in.	about ____ in.	about ____ in.
around wrist	about ____ in.	about ____ in.	about ____ in.
waist to floor	about ____ in.	about ____ in.	about ____ in.
forearm	about ____ in.	about ____ in.	about ____ in.
hand span	about ____ in.	about ____ in.	about ____ in.
arm span	about ____ in.	about ____ in.	about ____ in.
_____	about ____ in.	about ____ in	about ____ in.
_____	about ____ in.	about ____ in.	about ____ in.
_____	about ____ in.	about ____ in.	about ____ in.

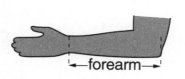
←forearm→

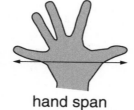
hand span

arm span

LESSON 3·4

Math Boxes

1. Count back by 7s.

_____ , _98_ , _____ , _____ ,

_____ , _____ , _63_ , _____ ,

_____ , _____ , _____ , _____ ,

_____ , _____ , _____ , _____

2. Measure to the nearest centimeter.

about _____ cm

_____

Draw a line segment 4 centimeters long.

SRB 137–139

3. Write <, >, or =.

1,069 _____ 10,691

6,589 _____ 6,859

42,617 _____ 42,429

Make up your own.

_____ _____ _____

SRB 13

4. 53 people were standing in line at 9:00 A.M. 97 people were standing in line at 10:00 A.M. How many more people were standing in line at 10:00 A.M.?

_____ people

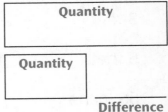
Quantity

Quantity

Difference

SRB 258

5. Shade to show the following data:
A is 4 cm.
B is 3 cm.
C is 8 cm.
D is 7 cm.

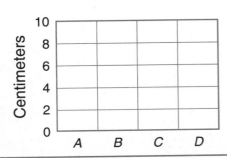

SRB 86

6. 2 children share 12 toys equally. How many toys does each child get?

_____ toys

SRB 73

Math Boxes

1. Measure to the nearest $\frac{1}{2}$ inch. Fill in the oval next to the best answer.

_____

  ⬭ 1 in.

  ⬭ $1\frac{1}{2}$ in.

  ⬭ 2 in.

  ⬭ $2\frac{1}{2}$ in.

SRB
143 144

2. What is the perimeter?

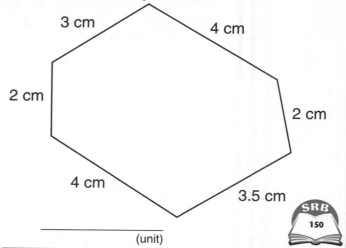

3 cm   4 cm

2 cm   2 cm

4 cm   3.5 cm

_____ (unit)

SRB
150

3. Write <, >, or =. Use a tape measure to help.

$1\frac{1}{2}$ feet _____ 16 inches

3 feet _____ 2 yards

5 feet _____ 60 inches

55 inches _____ 1 yard

SRB
13, 146

4. Add. Show your work.

Ballpark estimate:

Unit

_____

$$\begin{array}{r} 555 \\ +\ 192 \\ \hline \end{array}$$

SRB
57–59
192

5. Solve.

Unit

$9 + 1 + 4 =$ _____

_____ $= 3 + 7 + 8 + 2$

$3 + 15 + 7 + 4 =$ _____

SRB
50 51

6. Solve.

$3 \times 0 =$ _____

_____ $= 5 \times 0$

$0 \times 7 =$ _____

$9 \times 0 =$ _____

SRB
56

LESSON 3·6 Geoboard Perimeters

Materials ☐ geoboard and rubber bands or geoboard dot paper

Work with a partner.

1. Suppose that the distance between two pins is 1 unit. Make a rectangle with a perimeter of 14 units. Use rubber bands and a geoboard, or draw the rectangle on dot paper. Record the lengths of the sides in the table.

2. Now make a different rectangle that also has a perimeter of 14 units. Record the lengths of the sides for this shape.

3. Complete the table for other perimeters.

4. Try to make a rectangle or square with a perimeter of 13 units.

5. Try to make other rectangles or squares with perimeters that are an odd number of units.

Geoboard Perimeters		
Perimeter	Longer sides	Shorter sides
14 units	_____ units	_____ units
14 units	_____ units	_____ units
14 units	_____ units	_____ units
12 units	_____ units	_____ units
12 units	_____ units	_____ units
12 units	_____ units	_____ units
16 units	_____ units	_____ units
16 units	_____ units	_____ units
16 units	_____ units	_____ units
16 units	_____ units	_____ units

Try This

Change the unit. Now 1 unit is double the distance between two points. Make a rectangle or square whose perimeter is an odd number of units.

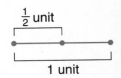

Follow-Up

Look for a pattern in your table. Can you find one? Now, without using a geoboard or dot paper, find the lengths of the sides of a rectangle or square with a perimeter of 24 units. Then make or draw the shape to check your answer.

LESSON 3·6 Tiling with Pattern Blocks

Materials ☐ pattern blocks: square, triangle, narrow rhombus
☐ crayons

Work with a partner.

1. Use square pattern blocks. Look at the top rectangle on the next page. Cover as much of the rectangle as you can, placing all of the blocks inside it. There may be uncovered spaces at the edges. Do not overlap the blocks. Line them up so that there are no gaps. This is called **tiling.**

2. Count and record the number of blocks you used.

3. Trace around the edges of each block. Then color any spaces not covered by blocks. Estimate how many blocks would be needed to cover the colored spaces.

4. Record how many blocks are needed to cover the whole rectangle.

5. Tile the second rectangle with triangles. Repeat Steps 2–4 above.

6. Tile the third rectangle with narrow rhombuses. Repeat Steps 2–4 above.

Follow-Up

7. The **area** of a shape is a measure of the space inside the shape. You measured the area of a rectangle three ways: with squares, triangles, and narrow rhombuses. Record the areas below.

The area of the rectangle is about _____ squares.

The area of the rectangle is about _____ triangles.

The area of the rectangle is about _____ narrow rhombuses.

8. Which of the three pattern blocks has the largest area? _____

Which has the smallest area? _____

How did you decide? _____

_____

Tiling with Pattern Blocks *continued*

Cover this rectangle with squares.

About _____ squares cover the whole rectangle.

Cover this rectangle with triangles.

About _____ triangles cover the whole rectangle.

Cover this rectangle with narrow rhombuses.

About _____ narrow rhombuses cover the whole rectangle.

LESSON 3·6 Straw Triangles

Materials ☐ 4-inch, 6-inch, and 8-inch straws

☐ twist-ties

Work in a group to make as many different-size triangles as you can out of the straws and twist-ties. (Be sure that straws are touching at all ends.) Before you start, decide how you will share the work.

For each triangle, record the length of each side and the perimeter in the chart. The triangle made out of the shortest straws is already recorded.

Straw Triangles			
Side 1	Side 2	Side 3	Perimeter
4 in.	4 in.	4 in.	12 in.

Follow-Up

Discuss these questions with others in your group.

1. Which triangles have three equal sides?

   _____

2. Which pairs of triangles have the same perimeter?

   _____

3. By looking at your constructions, estimate which triangle of each pair of triangles in Problem 2 has the larger area (space inside the triangles).

   _____

4. What happens if you try to make a triangle out of two 4-inch straws and one 8-inch straw?

   _____

70   seventy

LESSON 3·6

Math Boxes

---

1. Which tool would you use to measure the following?

| yardstick | ruler | thermometer |

temperature _____

height of the ceiling _____

length of your thumb _____

---

2. On the centimeter grid below, draw a shape with an area of 12 square centimeters.

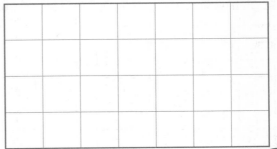

SRB 154 155

---

3. Justin bought 2 gallons of milk. Each gallon cost $2.79. He paid with a $10 bill. How much change did he receive? _____

SRB 250–253

---

4. Sarah had $0.54. She found some coins. Now she has $0.83. How much money did she find?

_____

Change

| Start | | End |

_____

SRB 254 255

---

5. Fill in the circle next to the best answer.

○ A. It is certain to be sunny tomorrow.

○ B. A tossed quarter will land on either HEADS or TAILS.

○ C. A rolled die will always land on 6.

○ D. The sun will set at the same time it did last month.

SRB 92

---

6. Fill in the empty frames.

Rule

×2

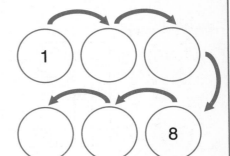

SRB 200 201

LESSON 3·7 Areas of Rectangles

Draw each rectangle on the grid. Make a dot inside each small square in your rectangle.

1. Draw a 3-by-5 rectangle.

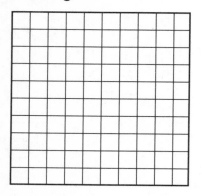

Area = _____ square units

2. Draw a 6-by-8 rectangle.

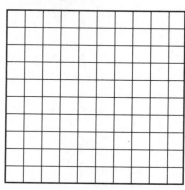

Area = _____ square units

3. Draw a 9-by-5 rectangle.

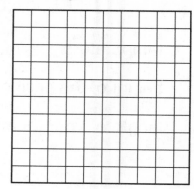

Area = _____ square units

Fill in the blanks.

4.

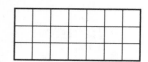

This is a _____-by-_____ rectangle.

Area = _____ square units

5.

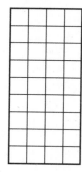

This is a _____-by-_____ rectangle.

Area = _____ square units

6.

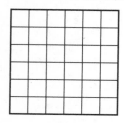

This is a _____-by-_____ rectangle.

Area = _____ square units

7.

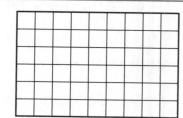

This is a _____-by-_____ rectangle.

Area = _____ square units

LESSON 3·7 — Math Boxes

1. Measure to the nearest centimeter.

about _____ cm

_____

Draw a line segment 4 centimeters long.

SRB
137–139

2. Draw a shape with an area of 9 square centimeters.

SRB
154   155

3. Write the equivalent lengths. Use a tape measure to help.

3 yards = _____ feet

_____ inches = 2 yards

50 millimeters = _____ centimeters

3 meters = _____ centimeters

SRB
140   146

4. Subtract. Show your work.

Ballpark estimate:

_____

$$943 - 409$$

Unit

SRB
60–63
192

5. Solve.

Unit

8 + 3 + 2 + 2 = _____

_____ = 9 + 14 + 1 + 2

4 + 3 + 11 + 6 = _____

85 + 16 + 4 + 15 = _____

SRB
50   51

6. Solve.

1 × 2 = _____

_____ = 1 × 4

5 × 1 = _____

8 × 1 = _____

SRB
52   53
56

LESSON 3·8 More Areas of Rectangles

Make a dot inside each small square in one row. Then fill in the blanks.

1.

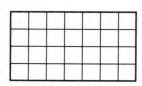

Number of rows: _____

Squares in a row: _____

Area = _____ square
units

Number Model:

_____ ✕ _____ = _____

2.

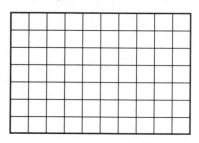

Number of rows: _____

Squares in a row: _____

Area = _____ square
units

Number Model:

_____ ✕ _____ = _____

3.

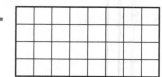

Number of rows: _____

Squares in a row: _____

Area = _____ square
units

Number Model:

_____ ✕ _____ = _____

Now draw the rectangle on the grid. Then fill in the blanks.

4. Draw a 5-by-7
rectangle.

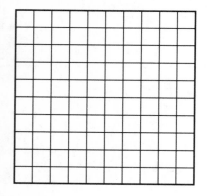

Area = _____ square
units

Number Model:

_____ ✕ _____ = _____

5. Draw an 8-by-8
rectangle.

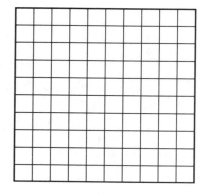

Area = _____ square
units

Number Model:

_____ ✕ _____ = _____

6. Draw a 3-by-9
rectangle.

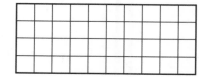

Area = _____ square
units

Number Model:

_____ ✕ _____ = _____

LESSON 3·8 Math Boxes

---

1. Circle the best unit of measurement.

distance to Spain

miles   centimeters   inches

width of a crayon

miles   centimeters   feet

length of your journal

miles   yards   inches

SRB 141 142 148 149

2. Find the perimeter.

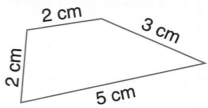

2 cm

3 cm

2 cm

5 cm

Fill in the circle for the best answer.

○ A. 13 cm

○ B. 12 cm

○ C. 8 cm

○ D. 4 cm

SRB 150 151

---

3. When I left home, I had $4.00. I spent $0.73 at the fruit stand and $2.59 at the grocery store. How much did I spend in all?

_____

How much do I have left?

_____

SRB 250–253

4. Fill in the unit box. Write the missing number in the diagram.

Unit
_____

Change

Start		End
	+107	392

Write a number model.

_____ + _____ = _____

SRB 254 255

---

5. Describe 2 events that you are certain *will not* happen today.

_____

_____

_____

SRB 92

6. Fill in the empty frames.

Rule

×0

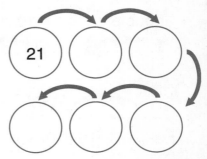

21

SRB 56 200 201

LESSON 3·9   # Diameters and Circumferences

1. Find numbers on the label of your can. Write some of them below. Also write the unit if there is one.

   _____

2. Record the diameter and circumference of your can.

   can letter: _____   **diameter:** about _____ cm

   circumference: about _____ cm

3. Write the rule linking diameter and circumference:

   _____

   _____

4. Fill in the empty frames. Use two rules.

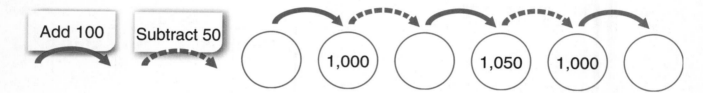

5.

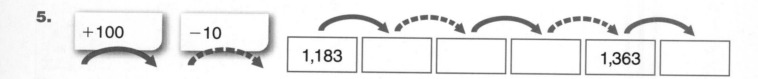

6.

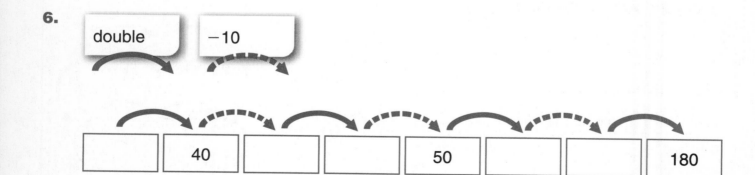

LESSON 3·9   Math Boxes

1. Measure to the nearest centimeter.

about _____ cm

_____

Draw a line segment 7 centimeters long.

SRB
137–139

2.

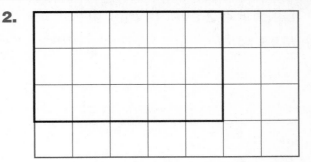

Area: _____ square cm

SRB
154 155

3. Write <, >, or =. You may wish to use a tape measure.

6 decimeters _____ 60 millimeters

3 yards _____ 36 inches

2 centimeters _____ 4 meters

Write your own.

_____

SRB
13 140
146

4. Add. Show your work.   **Unit**

Ballpark estimate:

_____ + _____ = _____

8,916
+ 7,504
_____

SRB
57 59
192

5. Write <, >, or =.   **Unit**

4 + 5 + 6 _____ 3 + 5 + 7

7 + 9 + 5 _____ 6 + 6 + 8

2 + 11 + 4 _____ 7 + 1 + 9

15 + 7 + 5 _____ 9 + 9 + 9

4 + 5 + 6 _____ 3 + 7 + 6

SRB
13
50 51

6. Solve.

$2 \times 2 =$ _____

$2 \times 3 =$ _____

_____ $= 2 \times 4$

$2 \times 6 =$ _____

SRB
52 53

Math Boxes

---

1. Describe 2 events that are *impossible*.

_____

_____

_____

SRB
92

2. Karan shared 18 jelly beans equally with her sister Sonia. How many jelly beans did they each get? Draw a picture or an array.

_____ jelly beans

SRB
73

---

3. Solve.

$4 \times 0 =$ _____

$0 \times 3 =$ _____

_____ $= 8 \times 0$

_____ $= 0 \times 10$

SRB
56

4. Solve.

$3 \times 1 =$ _____

_____ $= 1 \times 7$

_____ $= 4 \times 1$

$1 \times 6 =$ _____

SRB
56

---

5. Solve.

$2 \times 5 =$ _____

$1 \times 2 =$ _____

_____ $= 10 \times 2$

_____ $= 2 \times 7$

SRB
52 53

6. Fill in the empty frames.

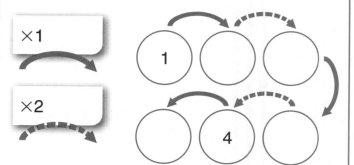

SRB
200 201
52 53

LESSON 4·1 | Solving Multiplication Number Stories

Use the Variety Store Poster on page 215 of the *Student Reference Book.*

For each number story:

◆ Fill in a multiplication/division diagram. Write ? for the number you need to find. Write the numbers you already know.

◆ Use counters or draw pictures to help you find the answer.

◆ Record the answer with its unit. Check whether your answer makes sense.

1. Yosh has 4 boxes of mini stock cars. There are 10 stock cars in each box. How many stock cars does he have?

boxes	cars per box	cars in all

Answer: _____
                (unit)

How do you know your answer makes sense? _____

_____

2. There are 100 file cards in each package. How many cards are in 5 packages of file cards?

packages	cards per package	cards in all

Answer: _____
                (unit)

How do you know your answer makes sense? _____

_____

3. Use a separate sheet of paper. Write your own multiplication number story. Write how you know your answer makes sense.

LESSON 4·1 **Math Boxes**

---

1. Each square equals 1 sq cm. Find the area.

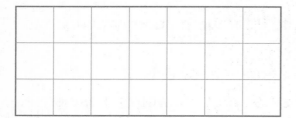

Area: _____ square centimeters

154–156

2. Scientists counted 136 eggs in a clutch of turtle eggs. 87 eggs did *not* hatch.

Estimate how many eggs hatched.

About _____

Number model for estimate:

_____

191

---

3. Make each sentence true. Use >, <, or =.

8 + 6 _____ 7 + 7

17 − 9 _____ 9 + 8

14 − 5 _____ 11 − 4

13
50 51

4. Find the perimeter.

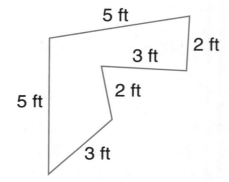

5 ft

3 ft   2 ft

5 ft

2 ft

3 ft

Perimeter = _____
(unit)

150 151

---

5. How many rows of dots?   • • • • •
                                  • • • •

_____

How many dots in each row?

_____

How many dots in all? _____

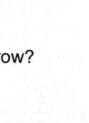
64 65

6. Solve.

45,582 − 100   = _____

45,582 + 100   = _____

45,582 + 1,000 = _____

45,582 − 1,000 = _____

18 19

LESSON 4·2 More Multiplication Number Stories

◆ Fill in the multiplication/division diagram.

◆ Make an array with counters. Mark the dots to show the array.

◆ Find the answer. Write the unit with your answer. Write a number model.

1. Mrs. Kwan has 3 boxes of scented markers. Each box has 8 markers. How many markers does she have?

boxes	markers per box	markers in all

Answer: _____
(unit)

Number model: _____

2. Monica keeps her doll collection in a case with 5 shelves. On each shelf there are 6 dolls. How many dolls are in Monica's collection?

shelves	dolls per shelf	dolls in all

Answer: _____
(unit)

Number model: _____

3. During the summer Jack mows lawns. He can mow 4 lawns per day. How many lawns can he mow in 7 days?

days	lawns per day	lawns in all

Answer: _____
(unit)

Number model: _____

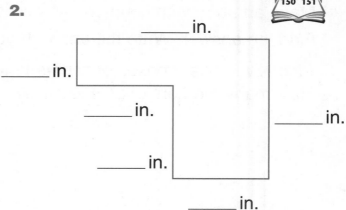

LESSON 4·2 Measuring Perimeter

Measure the perimeter of each figure in inches.

1.

_____ in.

_____ in.   _____ in.

_____ in.

Perimeter: _____ inches

2.

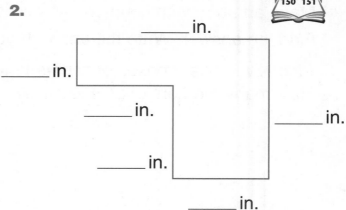

_____ in.

_____ in.

_____ in.

_____ in.

_____ in.

_____ in.

Perimeter: _____ inches

3.

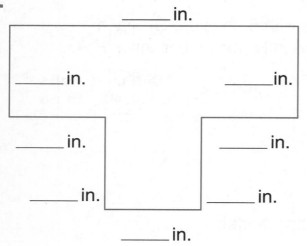

_____ in.

_____ in.   _____ in.

_____ in.   _____ in.

_____ in.   _____ in.

_____ in.

Perimeter: _____ inches

4. **Try This**

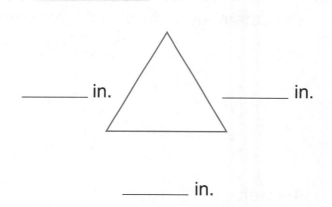

_____ in.   _____ in.

_____ in.

Perimeter: _____ inches

5. Draw any figure with a perimeter of 20 centimeters.

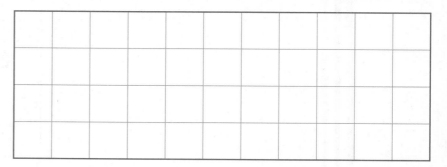

LESSON 4·2 — Math Boxes

1. Complete the bar graph.

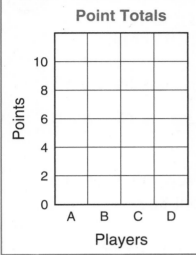

Point Totals

Player A scores 4 points.

Player B scores 8 points.

Player C scores 3 points.

Player D scores 9 points.

SRB
86 87

2. 10 packs of gum on the shelf in the candy store. 8 sticks of gum per pack. How many sticks of gum in all?

packs	sticks of gum per pack	sticks of gum in all

Answer: _____

SRB
259

3. Solve. Make a ballpark estimate to check that the answer makes sense.

Unit

_____ = 648 + 209

estimate:

_____

SRB
192

4. Solve.

3 × 5 = _____

3 nickels = _____ ¢

_____ = 4 × 5

_____ ¢ = 4 nickels

SRB
52 53
56

5. Fill in the empty frames.

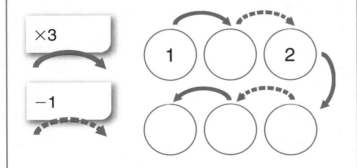

×3

−1

1   2

SRB
200 201

6. 1,798

Which digit is in the tens place? _____

Which digit is in the hundreds place? _____

Which digit is in the ones place? _____

Which digit is in the thousands place? _____

SRB
18 19

LESSON 4·3   Division Practice

Use counters to find the answers. Fill in the blanks.

16¢ shared equally

1. by 2 people:

_____¢ per person

_____¢ remaining

2. by 3 people:

_____¢ per person

_____¢ remaining

3. by 4 people:

_____¢ per person

_____¢ remaining

25¢ shared equally

4. How many people get 5¢?

_____ people

_____¢ remaining

5. How many people get 3¢?

_____ people

_____¢ remaining

6. How many people get 6¢?

_____ people

_____¢ remaining

30 stamps shared equally

7. by 10 people:

_____ stamps per person

_____ stamps remaining

8. by 5 people:

_____ stamps per person

_____ stamps remaining

9. by 6 people:

_____ stamps per person

_____ stamps remaining

10. 21 days
7 days per week

_____ weeks

_____ days remaining

11. 32 crayons
6 crayons per box

_____ boxes of crayons

_____ crayons remaining

12. 24 eggs
6 eggs per row

_____ rows of eggs

_____ eggs remaining

13. There are 18 counters in an array. There are 6 rows.

How many counters are in each row? _____ counters per row

14. Five children share 12 markers equally. How many markers does

each child get? _____ markers with _____ markers remaining

LESSON 4·3   **Math Boxes**

---

1. On the centimeter grid below, draw a shape with an area of 10 square centimeters.

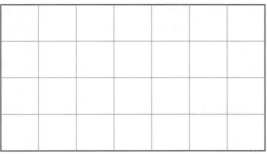

SRB 154–156

2. Corinne wants new tires for her bicycle. They cost $41.10 each, with tax included. Estimate about how much money she will need.

about $_____

Number model:

_____

SRB 191

---

3. Use >, <, or =.

Unit

$9 + 9$ _____ $13 + 5$

$13 - 4$ _____ $11 - 5$

$11 - 4$ _____ $13 - 8$

SRB 13 50 51

4. Find the perimeter. Fill in the circle for the best answer.

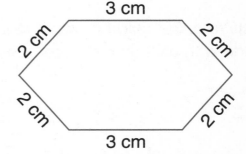

Ⓐ 14 cm   Ⓑ 6 cm

Ⓒ 7 cm   Ⓓ 12 cm

SRB 150 151

---

5. Complete the number model for the 4 by 4 array.

How many rows? _____

How many dots in each row?

_____

_____ × _____ = _____

SRB 64 65

6. Write the number that is 100 more.

76 _____

300 _____

471 _____

8,634 _____

5,925 _____

SRB 18 19

LESSON 4·4   **Solving Multiplication and Division Number Stories**

Solve each number story. Use counters or draw an array to help you. Fill in the diagrams and write number models.

SRB
253
259 260

1a. Roberto has 3 packages of pencils. There are 12 pencils in each package. How many pencils does Roberto have in all?

Answer: _____
            (unit)

Number model: _____

packages	pencils per package	pencils in all

1b. Roberto gives 3 of his pencils to each of his friends. How many friends will get 3 pencils each?

Answer: _____
            (unit)

Number model: _____

friends	pencils per friend	pencils in all

2a. A class of 30 children wants to play ball. How many teams can be made with exactly 6 children on each team?

Answer: _____
            (unit)

Number model: _____

teams	children per team	children in all

2b. For another game, the same class of 30 children wants to have exactly 4 children on each team. How many teams can they make?

Answer: _____
            (unit)

Number model: _____

teams	children per team	children in all

Math Boxes

1. Complete the bar graph.

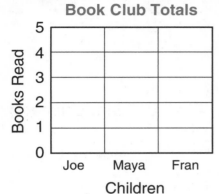

Book Club Totals

Books Read / Children

Joe read 3 books.

Maya read 2 books.

Fran read 4 books.

Median books read: _____ (unit)

SRB 80 86 87

2. Show an array and complete a number model to match the diagram.

packs	cards per pack	cards in all
3	6	?

Number model: _____

SRB 64 65 259 260

3. Add. Make a ballpark estimate.

Unit

_____ = 47 + 192

estimate: _____

_____ = 147 + 292

estimate: _____

SRB 57–59 192

4. Solve.

$6 \times 10 =$ _____

6 dimes = _____ ¢

_____ $= 8 \times 10$

_____ ¢ = 8 dimes

SRB 52 53 56

5. Fill in the empty frames.

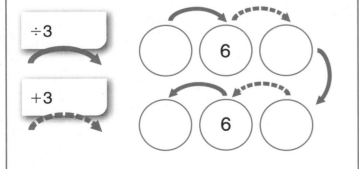

÷3

+3

6

6

SRB 200 201

6. What number has

2 hundreds   5 ones   3 tens

4 thousands   6 ten-thousands

Fill in the circle for the best answer and read it to a partner.

(A) 24,536   (B) 42,356

(C) 63,542   (D) 64,235

SRB 18 19

LESSON 4·5 Subtraction Strategies

Make a ballpark estimate. Write a number model to show your estimate. Choose at least two problems to solve using the counting-up method. You may choose any method you wish to solve the other problems.

1. Ballpark estimate:

_____

$$\begin{array}{r} 226 \\ -134 \\ \hline \end{array}$$

2. Ballpark estimate:

_____

$$\begin{array}{r} 93 \\ -47 \\ \hline \end{array}$$

3. Ballpark estimate:

_____

$$\begin{array}{r} 487 \\ -129 \\ \hline \end{array}$$

4. Ballpark estimate:

_____

$$\begin{array}{r} 361 \\ -248 \\ \hline \end{array}$$

5. Ballpark estimate:

_____

$$\begin{array}{r} 724 \\ -396 \\ \hline \end{array}$$

6. Ballpark estimate:

_____

$$\begin{array}{r} 515 \\ -367 \\ \hline \end{array}$$

LESSON 4·5

Math Boxes

1. Use the dots to show a 3 × 6 array.

. . . .
. . . .
. . . .
. . . .
. . . .

What is the number model?

_____ × _____ = _____

SRB 64 65

2. Maximum number of points scored:

_____

Minimum number of points scored:

_____

Range of points scored:

_____

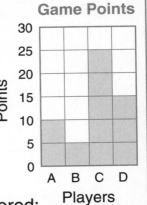

Game Points

SRB 79 86 87

3. Solve. Fill in the oval for the best answer.

4 rows of chairs

6 chairs in each row

How many chairs in all?

◯ 10 chairs   ◯ 12 chairs

◯ 24 chairs   ◯ 20 chairs

SRB 66 67

4. Fill in the number grid.

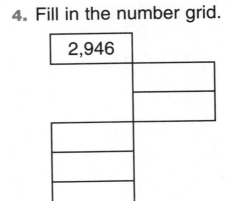

2,946

SRB 7–9

5. Draw a 2 × 4 rectangle.

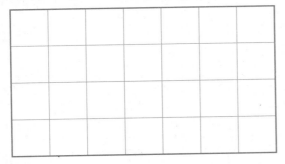

Number model: ___ × ___ = ___

Area: ___ square units

SRB 154–156

6. Which is more?

$3.45 or $3.09? _____

$0.34 or $0.09? _____

$14.50 or $14.55? _____

$30.15 or $31.05? _____

SRB 36

LESSON 4·6   Math Boxes

1. Write <, >, or =.

3×2 _____ 2×3

4×1 _____ 8×0

5×3 _____ 5×4

9×0 _____ 0×7

SRB 13 56

2. Write the fact family.

_____

_____

_____

_____

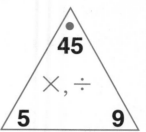

SRB 55

3. Complete.

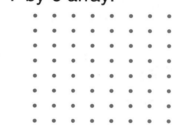

yd	ft
2	
5	
	9
	30

Rule
×3

SRB 52 203 204

4. Use the dots to show a 7-by-6 array.

. . . .
. . . .
. . . .
. . . .
. . . .
. . . .
. . . .

What is the number model?

_____ × _____ = _____

SRB 64 65

5. Lengths (in.) of 13 cats, including the tails:

30, 29, 28, 24, 29, 35, 16, 27, 29, 36, 28, 31, 32

What is the maximum length? Fill in the oval for the best answer.

◯ 36 inches   ◯ 16 inches

◯ 29 inches   ◯ 30 inches

SRB 79

6. Fill in the empty frames.

Rule
+1,000

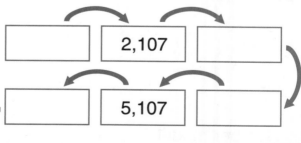

2,107

5,107

SRB 200 201

Math Boxes

1. Draw Xs in a 5-by-9 array.

How many Xs?

_____

Write a number model for the array.

_____

SRB
64 65

2. Maximum height of seedlings:

_____

Minimum height of seedlings:

_____

Range of seedling heights:

_____

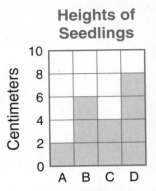
Heights of Seedlings

Centimeters
10
8
6
4
2
0
A B C D
Seedlings

SRB
79 86

3. Use counters to solve.

Some children are sharing 22 marbles equally. Each child gets 6 marbles. How many children are sharing?

_____
(unit)

How many marbles are left over?

_____
(unit)

SRB
73

4. Fill in the number grid.

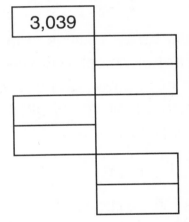
3,039

SRB
7–9

5. Draw a shape with an area of 11 square centimeters.

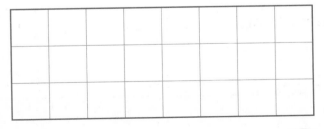

SRB
154–156

6. Use >, <, or =.

$40.75 _____ $47.05

$0.86 _____ $8.00

$31.02 _____ $31.20

$107.40 _____ $97.40

SRB
13 36

LESSON 4·8 Exploration A: How Many Dots?

Materials   ☐ square pattern blocks

☐ calculator

1. Estimate how many dots are in the array at the right.

   About _____ dots

   Make another estimate. Follow these steps.

2. Cover part of the array with a square pattern block. About how many dots does one block cover?

   _____ dots

3. Cover the array. Use as many square pattern blocks as you can. Do not go over the borders of the array. How many blocks did you use?

   _____ blocks

4. Use the information in Steps 2 and 3 to estimate the total number of dots in the array. About _____ dots

Try This

5. Find the exact number of dots in the array. Use a calculator to help you. Total number of dots = _____

Follow-Up

Describe how you found the exact number of dots. _____

_____

_____

LESSON 4·8

Exploration B: Setting Up Chairs

1. Record the answer to the problem about setting up chairs from *Math Masters,* page 106.

   There were _____ chairs in the room.

2. Circle dots below to show how you set up the chairs for each of the clues.

Rows of 2	Rows of 3	Rows of 4	Rows of 5
• •	• • •	• • • •	• • • • •
• •	• • •	• • • •	• • • • •
• •	• • •	• • • •	• • • • •
• •	• • •	• • • •	• • • • •
• •	• • •	• • • •	• • • • •
• •	• • •	• • • •	• • • • •
• •	• • •	• • • •	• • • • •
• •	• • •	• • • •	• • • • •
• •	• • •	• • • •	• • • • •
• •	• • •	• • • •	• • • • •
• •	• • •	• • • •	• • • • •
• •	• • •	• • • •	• • • • •
• •	• • •	• • • •	• • • • •
• •	• • •	• • • •	• • • • •
• •	• • •	• • • •	• • • • •
• 1 left over	• 1 left over	• 1 left over	• 0 left over

LESSON 4·8   Math Boxes

1. Make equal groups.

30 days make _____ weeks

with _____ days left over.

56 pennies make _____ quarters

with _____ pennies left over.

SRB 73

2. Write three things that you think are *very likely* to happen.

_____

_____

_____

_____

3. Fill in the circle for the best answer. The perimeter of the square is

○ A. 12 cm

○ B. 16 cm

○ C. 8 cm

○ D. 20 cm

4 cm

4 cm

SRB 150 151

4. Complete the Fact Triangle. Write the fact family.

_____ × _____ = _____

_____ × _____ = _____

_____ ÷ _____ = _____

_____ ÷ _____ = _____

21

×, ÷

3   _____

SRB 55

5. 56,937

Which digit is in the tens place? _3_

Which digit is in the thousands place? ___

Which digit is in the hundreds place? ___

Which digit is in the ones place? ___

SRB 18 19

6. Use >, <, or =.

3,065 _____ 3,605

23,605 _____ 20,365

32,605 _____ 23,605

50,007 _____ 50,700

SRB 13 20

LESSON 4·9 Estimating Distances

Locations to Visit

1. Mount St. Helens
2. Theme Park
3. Yellowstone National Park
4. Pikes Peak
5. Sears Tower

6. Civil Rights Memorial
7. Statue of Liberty
8. White House
9. Cowboy Hall of Fame
10. Space Shuttle Launch Site, Cape Canaveral

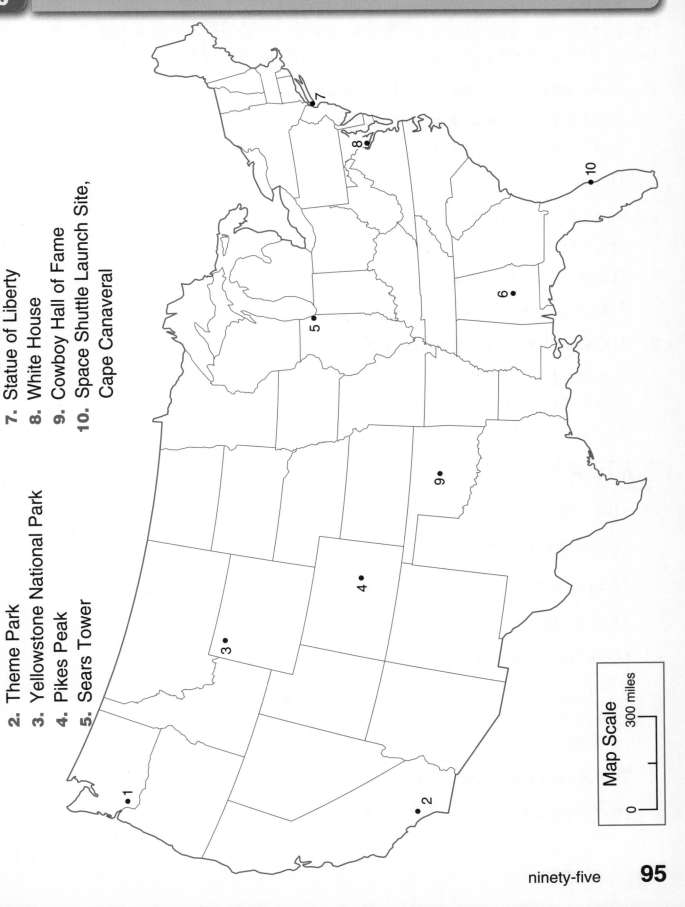

Map Scale

0   300 miles

LESSON 4·9   A Pretend Trip

Pretend that you want to take a trip to see some of the sights in the
United States. Find out about how far it is between locations.

1. Yellowstone National Park is number _____.

    The Cowboy Hall of Fame is number _____.

    The distance between them is about _____ inches on the map.

    That is about _____ miles.

2. Pikes Peak is number _____.

    The White House is number _____.

    The distance between them is about _____ inches on the map.

    That is about _____ miles.

3. The Civil Rights Memorial is number _____.

    Disneyland is number _____.

    The distance between them is about _____ inches on the map.

    That is about _____ miles.

Try This

4. The Statue of Liberty is number _____.

    The Sears Tower is number _____.

    The distance between them is about _____ inches on the map.

    That is about _____ miles.

5. Make up one of your own.

    _____ is number _____.

    _____ is number _____.

    The distance between them is about _____ inches on the map.

    That is about _____ miles.

LESSON 4·9 **Math Boxes**

---

1. Write $<$, $>$, or $=$.

5×5 _____ 4×4

1×9 _____ 7×1

6×2 _____ 2×8

3×7 _____ 7×3

SRB 13 56

---

2. Complete the Fact Triangle and write the fact family.

____ $\times$ ____ = ____

____ $\times$ ____ = ____

____ $\div$ ____ = ____

____ $\div$ ____ = ____

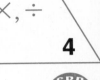

$\times$, $\div$

7   4

SRB 55

---

3. Complete.

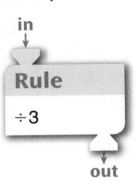

Rule

$\div 3$

in	out
9	
15	
	7
	10

SRB 52 203–204

---

4. Draw an array of 28 Xs arranged in 4 rows.

How many Xs in each row? _____

Write a number model for the array.

_____

SRB 64 65

---

5. Ages of 9 grandfathers:

60, 54, 79, 80, 65, 74, 70, 65, 81

mode = _____

median = _____

SRB 80 81

---

6. Fill in the empty frames.

Rule

$+1,000$

| | 5,670 | |

| | 8,670 | |

SRB 200 201

LESSON 4·10 Coin-Toss Experiment

Work with a partner. You need 10 coins.

1. You will each toss all 10 coins 5 times.

   For each toss you make,
   record the number of HEADS
   and the number of TAILS in the table.

Toss Record		
Toss (10 coins)	HEADS	TAILS
1		
2		
3		
4		
5		
Total		

2. Use the information in both your partner's and your tables to fill in the blanks below.

   My total: HEADS _____   TAILS _____

   My partner's total: HEADS _____   TAILS _____

   Our partnership total: HEADS _____   TAILS _____

3. Record the number of HEADS and the number of TAILS for the whole class.

   Number of HEADS: _____   Number of TAILS: _____

4. Suppose a jar contains 1,000 pennies. The jar is turned over. The pennies are
   dumped onto a table and spread out. Write your best guess for the number of
   HEADS and TAILS.

   Number of HEADS: _____   Number of TAILS: _____

LESSON 4·10   Measuring Line Segments

Use your ruler to measure each line segment.

Measure to the nearest $\frac{1}{2}$ inch.

1. _____

   about _____ inches

2. _____

   about _____ inches

3. _____

   about _____ inches

Try This

Measure to the nearest $\frac{1}{4}$ inch.

4. _____

   about _____ inches

5. _____

   about _____ inches

Measure to the nearest $\frac{1}{8}$ inch.

6. _____

   about _____ inches

LESSON 4·10   Math Boxes

1. 18 marbles are shared equally.

Each child gets 5 marbles.

How many children are sharing?

Use counters to solve.

_____
(unit)

How many marbles are left over?

_____
(unit)

2. Is a 6-sided die more likely to land on an odd number or on an even number? Explain.

_____

_____

_____

_____

_____

3. What is the perimeter of the rectangle?

12 in. ☐
      22 in.

The perimeter is _____.
                    (unit)

4. Complete the Fact Triangle. Write the fact family.

____ × ____ = ____

____ × ____ = ____

____ ÷ ____ = ____

____ ÷ ____ = ____

24

×, ÷

4   _____

5. Write the number that has

5 hundreds
7 thousands
8 ones
4 tens
2 ten-thousands

_____

Read it to a partner.

6. Use >, <, or =.

5,001 _____ 1,005

55,001 _____ 51,005

55,001 _____ 55,100

505,105 _____ 505,105

LESSON 4·11 **Math Boxes**

1. In the number 38,642

the 4 means _____.

the 8 means _____.

the 6 means _____.

the 3 means _____.

SRB
18 19

2. Put these numbers in order from smallest to largest.

4,073   47,003   43,700   7,430

_____   smallest

_____

_____

_____   largest

SRB
20

3. Solve.

Unit

12,469 + 10 = _____

12,469 + 100 = _____

12,469 + 1,000 = _____

12,469 + 10,000 = _____

SRB
18 19

4. Which is more?

$8.21 or $8.07   _____

$0.07 or $0.48   _____

$16.42 or $16.40   _____

SRB
36

5. Solve.

Unit

```
6,000      400
  300        9
   20    8,000
+   8    +   30
```

SRB
57 58

6. Write the number that is 100 more.

76 _____

300 _____

471 _____

8,634 _____

5,925 _____

SRB
18 19

LESSON 5·1   Place-Value Review

Ten-Thousands	Thousands	Hundreds	Tens	Ones

Follow the steps to find each number in Problems 1 and 2.

1. Write 6 in the ones place.
Write 4 in the thousands place.
Write 9 in the hundreds place.
Write 0 in the tens place.
Write 1 in the ten-thousands place.

2. Write 6 in the tens place.
Write 4 in the ten-thousands place.
Write 9 in the ones place.
Write 0 in the hundreds place.
Write 1 in the thousands place.

_____ _____ _____ _____ _____   _____ _____ _____ _____ _____

3. Compare the two numbers you wrote in Problems 1 and 2.

Which is greater? _____

4. Complete.

The 9 in 4,965 stands for 9 ___*hundreds*___ or ___*900*___ .

The 7 in 87,629 stands for 7 _____ or _____ .

The 4 in 48,215 stands for 4 _____ or _____ .

The 0 in 72,601 stands for 0 _____ or _____ .

Continue the counts.

5. 4,707; 4,708; 4,709; _____; _____; _____

6. 7,697; 7,698; 7,699; _____; _____; _____

7. 903; 902; 901; _____; _____; _____

8. 6,004; 6,003; 6,002; _____; _____; _____

9. 47,265; 47,266; 47,267; _____; _____; _____

LESSON 5·1   Math Boxes

1. If a map scale shows that 1 inch represents 200 miles, then

2 inches represent _____ miles

3 inches represent _____ miles

5 inches represent _____ miles

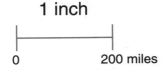

1 inch

0   200 miles

SRB
182

2. Put these numbers in order from smallest to largest.

54,752 _____

54,329 _____

54,999 _____

54,832 _____

SRB
18 19

3. Number of cookies in packages:

20, 24, 28, 30, 28, 26, 19, 24, 27

Put the data in order. Then find the median. Fill in the circle for the best answer.

Ⓐ 26 cookies   Ⓑ 25 cookies

Ⓒ 30 cookies   Ⓓ 1 cookie

SRB
80

4. Solve.

Unit

_____ = 80,000 − 40,000

_____ = 800,000 − 400,000

30,000 + 40,000 = _____

300,000 + 400,000 = _____

5.

JANUARY

Su	M	Tu	W	Th	F	Sa
15	16	17	18	19	20	21
		31				

January 17th is a Tuesday. What is the date on the following Tuesday?

_____

SRB
176 177

6. Use your template. Trace two different polygons.

SRB
102–105

LESSON 5·2 Math Boxes

1. Use multiplication or division to complete these problems on your calculator.

Enter	Change to	How?
10	5	÷ 2
3	15	
6	60	
45	5	

SRB 52 53

2. Maximum number of treats: _____

Minimum number of treats: _____

Range of number of treats: _____

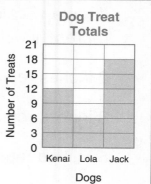

Dog Treat Totals

SRB 79

3. Find the total value. Fill in the circle for the best answer.

4   $1

3   Q

6   D

2   N

7   P

(A) $4.36

(B) $5.17

(C) $4.67

(D) $5.52

4. Barry exercises every day. He walked 11 laps on both Monday and Thursday, 8 laps on Tuesday, and 9 laps on Wednesday. How many laps did he walk in all?

_____
(unit)

Total			
Part	**Part**	**Part**	**Part**
11	11	8	9

 SRB 256 257

5. There are 3 cars. 4 people are riding in each car. How many people in all?

cars	people per car	people in all

Answer: _____
(unit)

Number model:

_____

SRB 259 260

6. Complete.

A triangle has _____ sides.

A rectangle has _____ sides.

A square has _____ sides.

SRB 106–109

LESSON 5·3   Math Boxes

---

1. If a map scale shows that 1 cm represents 1,000 km, then

2 cm represent _____ km

9 cm represent _____ km

16 cm represent _____ km

1 cm

0   1,000 km

182

---

2. Circle the largest number. Underline the smallest number.

946,487

946,800

946,793

946,200

SRB
18 19

---

3. Number of children in third grade classrooms:

31, 23, 21, 18, 28, 26, 22, 19, 30

What is the median?

_____ children

SRB
80

---

4. Solve.

Unit

$70,000 + 80,000 =$ _____

$700,000 + 800,000 =$ _____

_____ $= 12,000 - 5,000$

_____ $= 120,000 - 50,000$

---

5.   JANUARY

Su	M	Tu	W	Th	F	Sa
29	30	31				

If January 31st is on a Tuesday, what day of the week will it be on February 1st?

_____

176 177

---

6. Trace all the quadrangles on your template.

108 109

LESSON 5·4

Working with Populations

Populations of 10 U.S. Cities		
City	1990*	2000*
New York, NY	7,322,564	8,008,278
Chicago, IL	2,783,726	2,896,016
Houston, TX	1,630,553	1,953,631
Philadelphia, PA	1,585,577	1,517,550
Phoenix, AZ	983,403	1,321,045
Detroit, MI	1,027,974	951,270
Baltimore, MD	736,014	651,154
Nashville, TN	510,784	569,891
Sacramento, CA	369,365	407,018
Montgomery, AL	187,106	201,568

*U.S. Census Data

Use this table to solve the problems.

1. List the cities that lost population from 1990 to 2000.

_____

2. Name a city where the population increased by more than 100,000.

_____

3. Which city had the greatest change in total population? _____

   By about how many people did the population change? _____

4. In 1990, which city had a population about half that of Detroit, Michigan?

_____

5. In 2000, which city had a population about double that of Montgomery, Alabama?

_____

6. In 2000, which two cities combined had a population about the same as that of Chicago, Illinois?

_____

LESSON 5·4 Math Boxes

1. Use multiplication or division to complete these problems on your calculator.

Enter	Change to	How?
6	24	× 4
24	3	
4	36	
36	6	

52 53

2. Here are the number of minutes 11 third graders spent doing homework: 25, 45, 55, 30, 35, 45, 60, 30, 45, 35, 40.

What is the mode? Fill in the circle with the best answer.

(A) 45 minutes (B) 25 minutes

(C) 60 minutes (D) 40 minutes

81

3. Complete.

20 dimes = $ _____

20 nickels = $ _____

20 quarters = $ _____

10 quarters and 7 dimes = $ _____

4. Travis jogged 20 minutes on Monday, 16 minutes on Tuesday, and 14 minutes on both Thursday and Friday. How many total minutes did he jog? _____
(unit)

Total			
Part	Part	Part	Part

256 257

5. 15 feet of ribbon. 3 feet in each yard of ribbon. How many yards of ribbon?

yards of ribbon	feet per yard	feet of ribbon in all

Answer: _____
Number model: (unit)

_____

259 260

6. Complete.

A pentagon has _____ sides.

A decagon has _____ sides.

An octagon has _____ sides.

103

LESSON 5·5 **How Old Am I?**

1. On what date were you born? _____

2. How old were you on your last birthday? _____ years old

3. About how many minutes old do you think you were on your last birthday? Mark an X next to your guess.

   _____ between 10,000 and 100,000 minutes

   _____ between 100,000 and 1,000,000 minutes

   _____ between 1,000,000 and 10,000,000 minutes

Use your calculator.

4. **a.** About how many days old were you on your last birthday? Do not include any leap year days.   _____

   b. That's about how many hours?   _____

   c. That's about how many minutes?   _____

Try This

Adding Leap Year Days

5. **a.** List all of the leap years from the time you were born to your last birthday.   _____

   b. That adds how many extra days to your last birthday?   _____

   c. How many extra minutes?   _____

6. Add the number of extra minutes to the number of minutes in your answer in Problem 4c. How many minutes are there in all?   _____

7. On my last birthday, I was about _____ minutes old.

LESSON 5·5   Math Boxes

1. Solve.

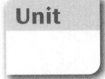
Unit

50,000 + 20,000 = _____

500,000 + 200,000 = _____

_____ = 90,000 − 50,000

_____ = 900,000 − 500,000

2. Write the number.

7 thousands

8 tens

5 ten-thousands

1 one

0 hundreds

4 hundred-thousands

___ ___ ___, ___ ___ ___

SRB
18–21

3. Write the following amounts using a dollar sign and decimal point:

2 dollar bills, 8 dimes, 9 pennies

_____

4 dimes, 6 pennies _____

1 dollar bill, 4 pennies _____

SRB
35

4. Solve using any method you wish.

Unit

$$\begin{array}{r} 907 \\ -\ 479 \\ \hline \end{array}$$

SRB
60–63

5. 13 crayons are shared equally among 3 children.

How many crayons does each child get?

_____
(unit)

How many crayons are left over?

_____
(unit)

SRB
259 260

6. Use a straightedge. Draw a polygon with 5 sides.

SRB
102–105

LESSON 5·6   Finding the Value of Base-10 Blocks

Exploration A:

Materials ☐ classroom supply of base-10 blocks

Work in a group.

1. Estimate the value of the base-10 blocks. Do not let anyone in your group see your estimate.

   Estimate: _____

2. Plan how your group will find the actual value of the blocks. Decide what each person will do to help. Carry out your plan.

3. What is the actual value of the base-10 blocks? _____

4. Write the estimates of your group and the actual value of the base-10 blocks in order from smallest to largest. Circle the actual value of the base-10 blocks.

   _____

   _____

5. Which estimate was closest to the actual value? _____

6. How many estimates were higher than the closest estimate? _____

7. How many estimates were lower than the closest estimate? _____

8. About how far was the highest estimate from the actual value? _____

9. About how far was the lowest estimate from the actual value? _____

10. How does your estimate compare to the actual value? _____

11. Describe how your group counted the blocks.

   _____

   _____

   _____

LESSON 5·6 Squares, Rectangles, and Triangles

Exploration B:

Materials ☐ straightedge

A
•

H •   • E

D •   •B

G •   •F

•
C

Work on your own or with a partner.

1. Use your straightedge to draw line segments between points
    A and *B*, *B* and *C*, *C* and *D*, and *D* and *A*.

    What kind of shape did you draw? _____

2. Now draw line segments between points *E* and *F*, *F* and *G*, *G* and *H*, and *H* and *E*.

    What kind of shape did you draw? _____

3. Draw line segments between points *E* and *G* and between points *F* and *H*.

    How many different sizes of squares are there? _____

    How many squares in all? _____

4. How many different sizes of triangles are there? _____

    How many triangles in all? _____

5. How many rectangles are there that are not squares? _____

LESSON 5·6 Pattern-Block Perimeters

Exploration C:

Materials ☐ pattern blocks: square, large rhombus, small rhombus, triangle

Work on your own or with a partner.

1. Imagine each polygon is rolled along a line, starting at point *S*. Estimate the distance each polygon will roll after 1 full turn. Mark an X at the point you think the polygon will reach.

2. Check your estimate by rolling a pattern block that matches the polygon. Circle the point reached by the pattern block.

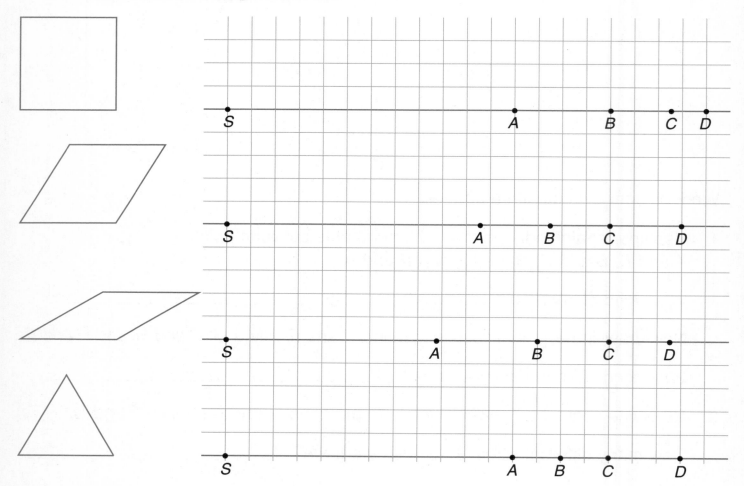

3. Which 3 shapes have about the same perimeter?

_____

4. Which of these 3 shapes has the largest area? _____

5. Which of the 4 shapes has the smallest area? _____

LESSON 5·6 Math Boxes

---

1. Circle the largest number.
Underline the smallest number.

56,689

86,953

90,865

65,398

SRB 18 19

---

2. Write the following amounts in dollars-and-cents notation.

5 dollar bills and 3 dimes _____

4 dollar bills and 6 dimes _____

1 dollar bill and 1 penny _____

SRB 35

---

3. The population of Chandler, Arizona, nearly doubled from 1990 to 2000. If 89,862 people lived in Chandler in 1990*, estimate about how many people lived there in 2000.

Ballpark estimate:

_____

Answer: _____
(unit)

*U.S. Census Bureau

SRB 191 193 194

---

4. Complete.

fishbowls	fish per bowl	fish in all
4	4	?

Answer: _____
(unit)

Number model: _____

SRB 259 260

---

5. Complete.

_____ days in a week

_____ days in two weeks

_____ days in three weeks

_____ days in four weeks

SRB 176 177

---

6. Measure each side of the quadrangle to the nearest half-centimeter.

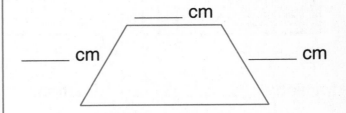

Another name for this quadrangle is a

_____ .
SRB 108 109 137–139

LESSON 5·7 Place Value in Decimals

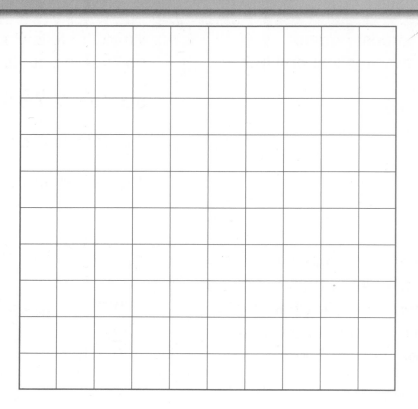

If the grid is ONE, then which part of each grid is shaded?

Write a fraction and a decimal below each grid.

1.

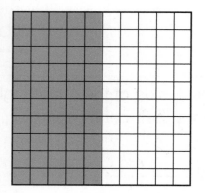

fraction: _____

decimal: _____

2.

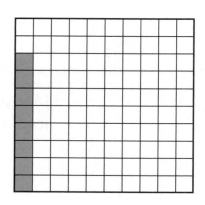

fraction: _____

decimal: _____

3.

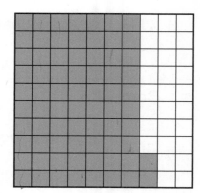

fraction: _____

decimal: _____

LESSON 5·7 **Place Value in Decimals** *continued*

4. Which decimal in each pair is greater? Use the grids in Exercises 1–3 to help you.

0.5 or 0.08 _____   0.08 or 0.72 _____   0.5 or 0.72 _____

Color part of each grid to show the decimal named.

5. Color 0.7 of the grid.

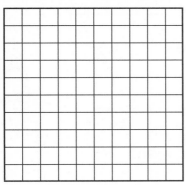

6. Color 0.07 of the grid.

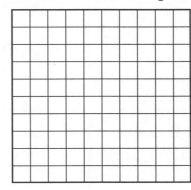

7. Color 0.46 of the grid.

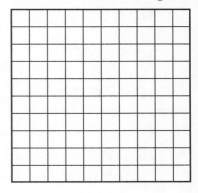

---

8. Write 0.7, 0.07, and 0.46 in order from smallest to largest.

Use the grids in Exercises 5–7 to help you. _____ _____ _____

Try This

Color part of each grid to show the fraction named.

9. Color $\frac{4}{10}$ of the grid.

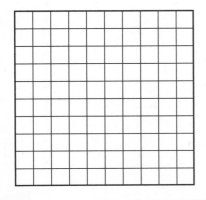

10. Color $\frac{1}{2}$ of the grid.

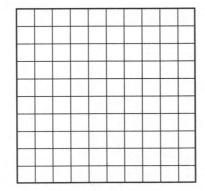

11. Color $\frac{23}{100}$ of the grid.

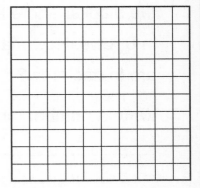

---

12. Write $\frac{23}{100}$ as a decimal. _____

LESSON 5·7   Math Boxes

1. Solve.

Unit

$16 + 9 =$ _____

$16 + 90 =$ _____

$16 + 900 =$ _____

$16 + 9,000 =$ _____

$16 + 90,000 =$ _____

2. For the number 5,749,862

the 4 means __40,000__

the 5 means _____

the 8 means _____

the 7 means _____

the 9 means _____

SRB 18–21

3. What is another name for 3 dollar bills and 6 pennies? Fill in the circle for the best answer.

Ⓐ $3.60

Ⓑ $3.06

Ⓒ $30.60

Ⓓ 3.06¢

SRB 35

4. Solve using any method you wish.

Unit

$$\begin{array}{r} 199 \\ +499 \\ \hline \end{array}$$

SRB 60–63

5. Write a division story by filling in the blanks. There are 48 _____ in 6 rows. How many _____ are in each row? _____

Write a number model.

_____

SRB 250–253 260

6. Use a straightedge. Draw a 7-sided polygon.

SRB 102–105

Exploring Decimals

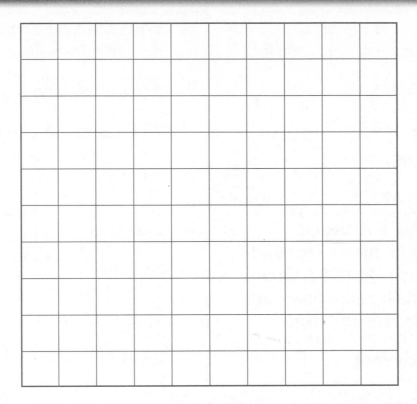

A	B	C	D
_13_ hundredths	_1_ tenth, _3_ hundredths	0. _13_	$\frac{13}{100}$
_____ hundredths	_____ tenths, _____ hundredths	0. _____	
_____ hundredths	_____ tenths, _____ hundredths	0. _____	
_____ hundredths	_____ tenths, _____ hundredths	0. _____	
_____ hundredths	_____ tenths, _____ hundredths	0. _____	
_____ hundredths	_____ tenths, _____ hundredths	0. _____	
_____ hundredths	_____ tenths, _____ hundredths	0. _____	
_____ hundredths	_____ tenths, _____ hundredths	0. _____	

LESSON 5·8 Math Boxes

---

1. Which number is the smallest? Fill in the circle for the best answer.

- ◯ **A.** 693,971
- ◯ **B.** 809,178
- ◯ **C.** 97,987
- ◯ **D.** 488,821

SRB 18 19

---

2. Write the following amounts in dollars-and-cents notation.

2 dollar bills and 3 dimes _____

7 dimes and 9 pennies _____

5 pennies _____

SRB 35

---

3. The land area of Alaska is 571,951 square miles. The land area of Texas is 261,797 square miles.* Estimate about how much bigger Alaska is than Texas.

Ballpark estimate:

_____

Answer: _____
(unit)

*data from *The World Almanac and Book of Facts 2004*

SRB 191 193 194

---

4. Complete.

tricycles	wheels per tricycle	wheels in all
?	3	24

Answer: _____
(unit)

Number model:

_____

SRB 259 260

---

5. Complete.

1 hour = _____ minutes

1 day = _____ hours

1 week = _____ days

1 year = _____ months

SRB 176 177

---

6. Measure each side of the polygon to the nearest half-inch.

_____ in. _____ in.

_____ in.

_____ in.

_____ in.

_____ in.

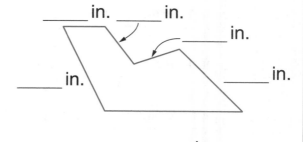

A 6-sided polygon is called a _____.

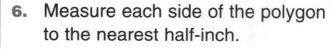

SRB 102 103 137–139

LESSON 5·9 — Decimals for Metric Measurements

1. Fill in the missing information. Put longs and cubes end to end on a meterstick to help you.

Length in Centimeters	Number of Longs	Number of Cubes	Length in Meters
24 cm	_2_	_4_	_0.24_ m
36 cm	_____	_____	_____ m
_____ cm	0	3	_____ m
8 cm	_____	_____	_____ m
_____ cm	_____	_____	0.3 m
_____ cm	4	3	_____ m

Work with a partner. Each partner uses base-10 blocks to make one length in each pair. Compare the lengths and circle the one that is greater.

2. 0.09 or 0.12

3. 0.24 or 0.42

4. 0.10 or 0.02

5. 0.18 or 0.5

6. 0.2 or 0.08

7. 0.3 or 0.24

Follow these directions on the ruler below. Use base-10 blocks and a meterstick to help you.

8. Make a dot at 4 cm and label it with the letter *A*.

9. Make a dot at 0.1 m and label it with the letter *B*.

10. Make a dot at 0.15 m and label it with the letter *C*.

11. Make a dot at 0.08 m and label it with the letter *D*.

LESSON 5·9   Math Boxes

1. Write the number that has

6 in the ones place
4 in the tenths place
3 in the hundredths place
2 in the thousandths place

____.____ ____ ____

SRB 35

2. If each grid is ONE, what part of each grid is shaded? Write the decimal.

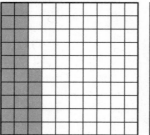
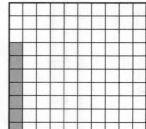

_____   _____

SRB 33–35

3. Use addition and subtraction to complete these problems on your calculator.

Enter	Change to	How?
894	12,894	
1,366	966	
627,581	628,581	
43,775	43,175	

SRB 18 19

4. Draw a 3 × 7 rectangle.

Number model: ____ × ____ = ____

Area: ____ square units

SRB 154–156

5. For the number 4,963,521

4 means _4,000,000_

3 means _____

1 means _____

6 means _____

9 means _____

SRB 18–21

6. Draw an example of a cylinder.

SRB 112–114

LESSON 5·10   How Wet? How Dry?

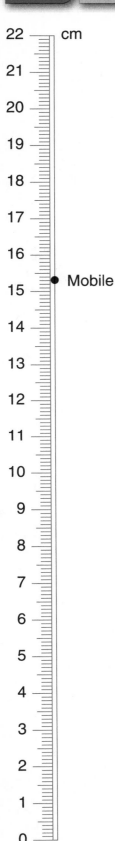

1. Use the scale at the left and the map on page 221 of the *Student Reference Book.* Make a dot for the level of precipitation in each of the following cities: Seattle, Omaha, Birmingham, and Tampa. Write the name of the city next to the dot.

2. Which city gets about 5 centimeters less rain than Mobile?

    _____

3. Which city gets about half as much rain as Omaha?

    _____

4. Which city gets about 4 times as much rain as Seattle?

    _____

5. A decimeter is 10 centimeters. Which cities on the map get at least 1 decimeter of rain?

    _____

Did You Know?

According to the National Geographic Society, the rainiest place in the world is Mount Waialeale in Hawaii. It rains an average of about 1,170 centimeters a year on Mount Waialeale.

Try This

6. Suppose it rained 1,170 centimeters in your classroom. Would the water reach the ceiling?

    _____ millimeters = 1,170 centimeters = _____ meters

    Answer: _____

LESSON
5·10   **Math Boxes**

---

1. Color 0.6 of the grid.

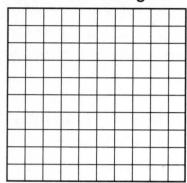

33 34

---

2. Complete the bar graph.

Lily ran
4 miles.

Meg ran
3 miles.

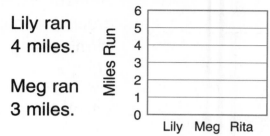

Rita ran
6 miles.

Median miles run: _____

86 87

---

3. Write <, >, or =.

0.65 _____ 0.56

0.07 _____ 0.7

0.098 _____ 0.102

73.4 _____ 75.2

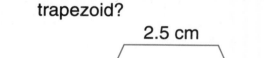
36

---

4. What is the perimeter of the trapezoid?

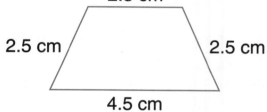

○ **A.** 10 cm   ○ **B.** 11 cm

○ **C.** 12 cm   ○ **D.** 13 cm

150 151

---

5. Write the number that has
2 in the ones place
6 in the tenths place
7 in the hundredths place

____.____ ____

35

---

6. This polygon has ____ sides.

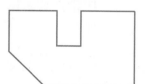

It is called a _____.

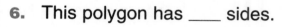

102 103

LESSON 5·11   More Decimals

Use your Place-Value Book to help you. Write the number that matches each description.

1. 0 in the ones place
    8 in the tenths place

    _____

2. 1 in the ones place
    3 in the tenths place

    _____

3. 2 in the ones place
    7 in the hundredths place
    0 in the tenths place

    _____

4. 0 in the hundredths place
    6 in the ones place
    8 in the thousandths place
    0 in the tenths place

    _____

5. Read each of the decimals in Problems 1–4 to a partner.

Write each number below as a decimal.

6. nine-tenths _____

7. thirty-thousandths _____

8. fifty-three hundredths _____

9. sixty and four-tenths _____

10. seven and seven-thousandths

    _____

11. sixty and four-hundredths

    _____

Unit
meter

Fill in the missing numbers.

12.

0 ___ ___ ___ ___ ___ ___ ___ ___ ___ 1

13.

0 ___ ___ ___ ___ ___ ___ ___ ___ ___ 0.1

LESSON 5·11 Math Boxes

1. Write the number that has

5 in the tenths place
4 in the hundredths place
1 in the ones place
6 in the thousandths place

____.____ ____ ____

SRB 35

2. How much of this grid is shaded? Fill in the circle for the best answer.

○ **A.** 0.58

○ **B.** 58

○ **C.** 5.8

○ **D.** 0.6

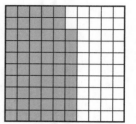

SRB 33–35

3. Use addition and subtraction to complete these problems on your calculator.

Enter	Change to	How?
629	18,629	
2,411	411	
456,972	450,972	
28,684	28,084	

SRB 18 19

4. Draw a shape with an area of 16 square units.

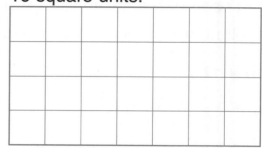

How many sides does your shape have?

_____ sides

SRB 154–156

5. For the number 3,975,081

5 means ___*5,000*___

1 means _____

7 means _____

9 means _____

3 means _____

SRB 18–21

6. An example of a sphere is a ball. Draw or name another sphere.

SRB 119

LESSON 5·12 Length-of-Day

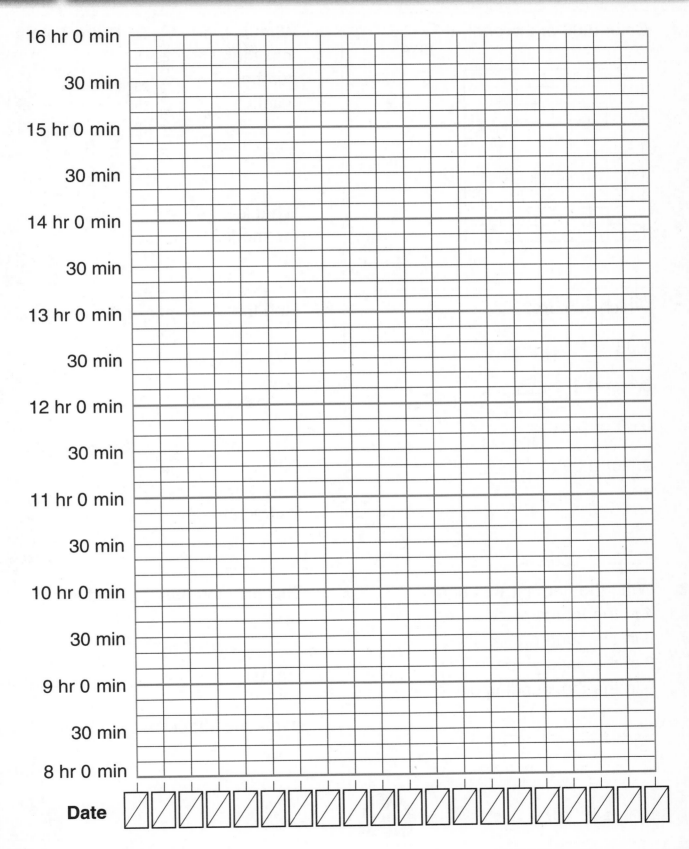

16 hr 0 min
30 min
15 hr 0 min
30 min
14 hr 0 min
30 min
13 hr 0 min
30 min
12 hr 0 min
30 min
11 hr 0 min
30 min
10 hr 0 min
30 min
9 hr 0 min
30 min
8 hr 0 min

Date

LESSON 5·12   Math Boxes

1. Color 0.08 of the grid.

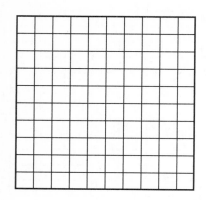

33 34

2. What is the maximum number of points?

Point Totals

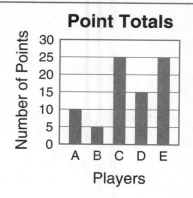

_____

What is the mode? _____

79 81 86

3. Which is more?

1.36 or 1.6 _____

0.4 or 0.372 _____

0.69 or 0.6 _____

0.7 or 0.09 _____

36

4. Find the perimeter of the octagon.

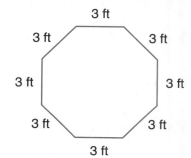

3 ft
3 ft   3 ft
3 ft   3 ft
3 ft   3 ft
3 ft

Perimeter =  _____
(unit)

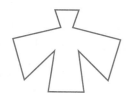
150 151

5. Write the number that has

4 in the tenths place
0 in the hundredths place
6 in the ones place
9 in the thousandths place

___ . ___ ___ ___

35

6. This polygon has ____ sides.

Name the shape. _____

102 103

LESSON 5·13 | **Math Boxes**

---

1. Draw a line that will divide the rectangle into 2 equal parts.

2. Complete.

A triangle has _____ sides

and _____ angles.

A quadrangle has _____

sides and _____ angles.

SRB
106–109

---

3. Draw a quadrangle.

SRB
108 109

4. Draw a polygon.

SRB
102 103

---

5. Use your template. Draw a rhombus and a square. How are they alike?

_____

_____

_____

6. Circle the pictures that show 3-dimensional shapes.

SRB
112–115

---

1. Write S next to each line segment. Write R next to each ray.
Write L next to each line.

  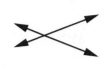 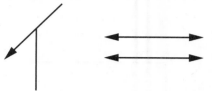

Points *D, T, Q,* and *M* are marked. Use a straightedge to draw the following.

2. Draw $\overline{QT}$. Draw $\overrightarrow{DT}$. Draw $\overleftrightarrow{MQ}$.

D
•

T
•

M
•

•*Q*

Draw a line segment between
each pair of points. How many line
segments did you draw?

Example:

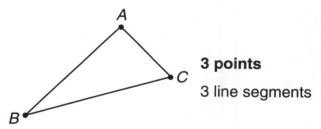

3 points

3 line segments

3.   *P*
                    •

  A •

          •*L*

      •
     U

4 points

_____ line segments

4.   *R*
                    •

  O •

                        •*E*

      S •

                    •*I*

5 points

_____ line segments

128   one hundred twenty-eight

LESSON 6·1 | **Math Boxes**

---

1. Complete the number-grid puzzle.

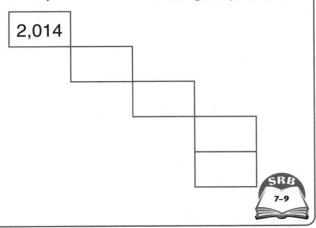

2,014

SRB 7–9

2. If the grid is ONE, then what part of the grid is shaded? Write a fraction and a decimal.

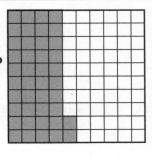

_____ = _____

(fraction)   (decimal)

SRB 33 34

---

3. What is a fair way to decide who should go first in a game?

_____

_____

_____

4. Cross out the names that do not belong in this name-collection box.

0.1	.01	$\frac{10}{100}$

$\frac{1}{100}$ one-hundredth

0.10   10

$\frac{1}{10}$ one-tenth

SRB 14 15 33

---

5. In the number 2.673,

the 6 means ___*6 tenths*___ .

the 3 means _____ .

the 7 means _____ .

the 2 means _____ .

SRB 35

6. Complete the Fact Triangle. Write the fact family.

_____

_____

_____

_____

72

×, ÷

8   ____

SRB 55

LESSON 6·2 Geometry Hunt

parallel line segments

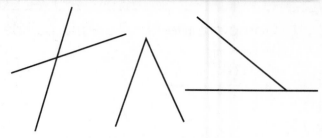

intersecting line segments

Part 1 (Use with Lesson 6-2.)

Look for things in the classroom or hallway that are parallel.
Look for things that intersect. List these things below or draw
a few of each of them on another sheet of paper.

Parallel

_____

_____

_____

Intersecting

_____

_____

_____

Part 2 (Use with Lesson 6-3.)

Look for things in the classroom or hallway that have one or more right angles.
List these things below or draw a few of them on another sheet of paper.

_____

_____

_____

Math Boxes

1. The grid is ONE.

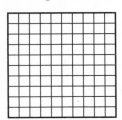

Shade 0.06 of the grid.   Shade 0.25 of the grid.

Write the larger number.

_____

SRB 34 36

2. 9 boxes of muffins. 6 muffins per box. How many muffins in all?

_____
                    (unit)

Write a number model:

_____ × _____ = _____

SRB 259 260

3. Use some or all of the cards to write different names for the target number.

3   2   5   4   6   12

target number

_____

_____

_____

_____

SRB 299 300

4.

Favorite Vegetables

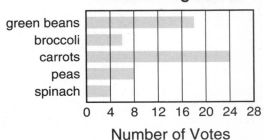

green beans
broccoli
carrots
peas
spinach

0 4 8 12 16 20 24 28
Number of Votes

Which vegetable is the least

favorite? _____

SRB 86 87

5. Draw a ray, $\overrightarrow{AB}$. Draw a line segment, $\overline{CD}$. Draw a line, $\overleftrightarrow{EF}$.

• A   • B

• C   • D

• E   • F

SRB 96 97

6. Solve.

Double 2 _____

Double 10 _____

Double 75 _____

Double 1,000 _____

Double 1,500 _____

LESSON 6·3   Turns

Use your connected straws to show each turn.
Draw a picture of what you did.
Draw a curved arrow to show the direction of the turn.

Example:

right $\frac{1}{4}$ turn (clockwise)	**1.** right $\frac{1}{2}$ turn (clockwise)	**2.** left $\frac{1}{4}$ turn (counterclockwise)
3. left $\frac{3}{4}$ turn (counterclockwise)	**4.** right $\frac{3}{4}$ turn (clockwise)	**5.** left $\frac{1}{2}$ turn (counterclockwise)

LESSON 6·3 **Math Boxes**

---

1. Complete the number-grid puzzle.

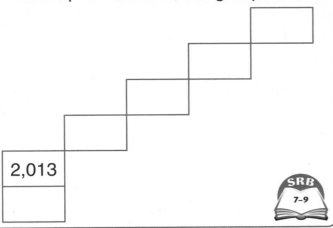

2,013

---

2. If the grid is ONE, then what part of the grid is shaded? Circle the best answer.

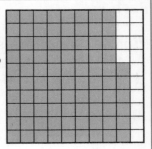

A   0.86   B   0.68

C   08.6   D   86

---

3. Lisa tosses a coin. How likely is the coin to land on HEADS? Circle one:

More likely to land on HEADS than on TAILS

Equally likely to land on HEADS or on TAILS

Less likely to land on HEADS than on TAILS

---

4. Cross out the names that do not belong in the name-collection box.

.05	$\frac{5}{10}$	5.0
$\frac{5}{100}$	five-tenths	0.05
five-hundredths		0.50

---

5. In the number 34.972,

the 9 means ___*9 tenths*___.

the 7 means _____.

the 3 means _____.

the 4 means _____.

the 2 means _____.

---

6. Complete the Fact Triangle. Write the fact family.

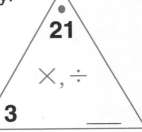

21

×, ÷

3   ___

_____

_____

_____

_____

LESSON 6·4 Exploring Triangles

Part 1

B •

Follow these steps:

1. Find the three points on the right.

2. Use a straightedge to connect each pair of points with a line segment.

3. What figure have you drawn?

•C

_____   •
                                                    A

Part 2

Write all six 3-letter names that are possible for your triangle. The first letter of each name is given below.

A_____   A_____   B_____   B_____   C_____   C_____

Part 3

Work with a group.

Make triangles with straws and twist-ties. Make at least one of each of the following kinds of triangles.

◆ all 3 sides the same length

◆ only 2 sides the same length

◆ no sides the same length

◆ 1 angle larger than a right angle

◆ all 3 angles smaller than a right angle

Part 4

Measure each side of the triangle you drew in Part 1 to the nearest $\frac{1}{4}$ inch.

side AB _____ in.   side BC _____ in.   side CA _____ in.

LESSON 6·4 Math Boxes

1.

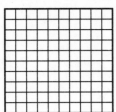

Shade 0.6 of the grid.   Shade 0.30.

Write the larger number. _____

34 36

2. 13 crayons are shared equally among 3 children. How many crayons does each child get?

Ⓐ 3 crayons, 4 left over

Ⓑ 4 crayons, 1 left over

Ⓒ 3 crayons, 1 left over

Ⓓ 4 crayons, 2 left over

73

3. Fill in the name-collection box with at least five equivalent names.

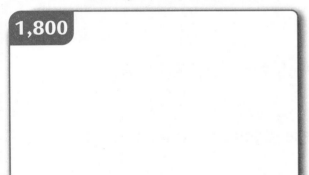

1,800

14 15

4. At what ages can the most girls arm hang for 8 seconds?

Arm Hanging for 8 Seconds

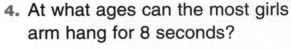
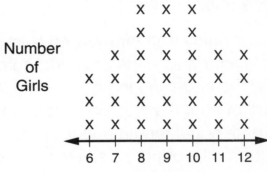

Number of Girls

```
                X  X  X
                X  X  X
          X  X  X  X  X  X
       X  X  X  X  X  X  X
       X  X  X  X  X  X  X
       X  X  X  X  X  X  X
    ◄──┼──┼──┼──┼──┼──┼──►
       6  7  8  9 10 11 12
```

Ages of Girls

77

_____

5. Draw $\overrightarrow{AT}$. Draw $\overline{BY}$. Draw $\overleftrightarrow{ME}$.

• A   • T

• B   • Y

• M   • E

96 97

6. Complete.

in	out
16	
	240
225	
133	
	1,000

in

Rule

double

out

203 204

LESSON 6·5 **Exploring Quadrangles**

Part 1

B •

Use a straightedge. Connect points to form
a quadrangle.

_____ A •

Part 2

Write all 4-letter names that are possible
for your quadrangle. The first letter of each   D •   • C
name is given below.

A_____   A_____   B_____   B_____

C_____   C_____   D_____   D_____

Part 3

Work in a group.

Make quadrangles with straws and twist-ties. Make at least one of
each of the following kinds of quadrangles.

- ◆ all 4 sides equal in length
- ◆ 2 pairs of equal-length sides, but opposite sides not equal in length
- ◆ 2 pairs of equal-length opposite sides
- ◆ only 2 parallel opposite sides
- ◆ only 1 pair of equal-length opposite sides

Part 4

Measure each side of the quadrangle you drew in Part 1 to the nearest
$\frac{1}{2}$ centimeter.

side AB _____ cm   side BC _____ cm   side CD _____ cm   side DA _____ cm

Try This

The perimeter of my quadrangle is about _____ centimeters.

LESSON 6·5   **Math Boxes**

---

1. The grid is ONE. Shade 0.41 of the grid.

Write the fraction that shows how much is shaded.

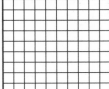

0.41 = _____

SRB 34

2. Circle the pair of lines that are parallel.

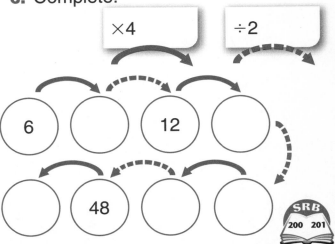

SRB 99

---

3. Fill in the oval for the best answer. The turn of the angle is

⬭ less than a $\frac{1}{2}$ turn.

⬭ less than a $\frac{1}{4}$ turn.

⬭ greater than a $\frac{1}{2}$ turn.

⬭ a full turn.

SRB 167 168

4. Draw a ray, $\overrightarrow{DO}$. Draw a line segment, $\overline{RE}$. Draw a line, $\overleftrightarrow{MI}$.

•   •

•   •

•   •

SRB 100

---

5. Draw a shape with 4 sides that are all equal in length.

This shape is a _____.

SRB 109

6. Complete.

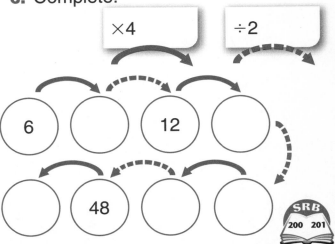

SRB 200 201

LESSON 6·6 Exploring Polygons

Part 1

A•

1. Use a straightedge and draw $\overline{AB}$, $\overline{BC}$, $\overline{CD}$, $\overline{DE}$, and $\overline{EA}$.

•B

E•

2. What kind of polygon did you draw?

   _____

•C

3. Write 4 or more possible letter names for the polygon.

D•

   _____   _____   _____

   _____   _____   _____

_____

Part 2

Work in a group to make polygons with straws and twist-ties. Your teacher will tell you how many sides each polygon should have.

Make at least one of each of the following kinds of polygons.

◆ all sides equal in length, and all angles equal in size (the amount of turn between sides)

◆ all sides equal in length but not all angles equal in size

◆ any polygon having the assigned number of sides

LESSON 6·6

Exploring Polygons *continued*

Part 3

A **regular polygon** is a polygon in which all the sides are equal and all the angles are equal.

Below, trace the smaller of each kind of *regular* polygon from your Pattern-Block Template.	Below, trace all the polygons from your Pattern-Block Template that are *not* regular polygons.

Part 4

Measure each side of the polygon you drew in Part 1 to the nearest $\frac{1}{2}$ centimeter.

side *AB* about _____ cm

side *BC* about _____ cm

side *CD* about _____ cm

side *DE* about _____ cm

side *EA* about _____ cm

Try This

The perimeter of my polygon is about _____ cm.

LESSON 6·6 — Math Boxes

1. A pentagon has

_____ sides,

_____ vertices,

and _____ angles.

Draw a pentagon.

SRB 103

2. If each grid is ONE, what part of each grid is shaded? Write the decimal.

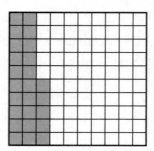 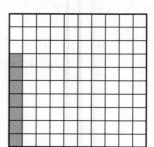

_____   _____

Circle the larger number.

SRB 34 36

3. Write <, >, or =.

0.45 _____ 0.54

1.07 _____ 1.7

2.3 _____ 0.23

10.8 _____ 10.80

0.2 _____ 2.0

SRB 36

4. 64 slices of pizza. 8 people. How many slices per person?

Fill in the oval for the best answer.

◯ 64 + 8
◯ 64 − 8
◯ 64 × 8
◯ 64 ÷ 8

SRB 73

5. Draw line segments to form a quadrangle.

M•   A•

•
H   •
        T

Which letter names the right angle?

_____

SRB 98 108 109

6. Complete the Fact Triangle. Write the fact family.

_____

_____

_____

_____

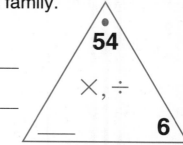

54

×, ÷

6

SRB 55

Drawing Angles

**LESSON
6·7**

Draw each angle as directed by your teacher.
Record the direction of each turn with a curved arrow.

Part 1

A •

B •

C •

Part 2

S •

R
•

T •

LESSON 6·7   **Math Boxes**

---

1. The grid is ONE. Shade $\frac{57}{100}$ of the grid.

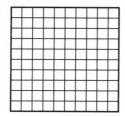

Write the decimal that tells how much of the grid is shaded.

_____

SRB
34

---

2. Circle the pairs of lines that intersect.

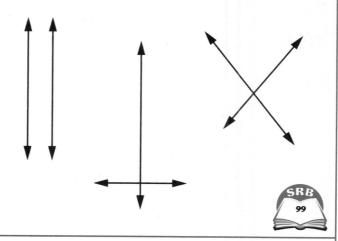

SRB
99

---

3. Draw an angle that is less than a $\frac{1}{4}$ turn.

SRB
167 168

---

4. Draw a ray, $\overrightarrow{SO}$. Draw a line segment, $\overline{LA}$. Draw a line, $\overleftrightarrow{TI}$.

SRB
100

---

5. Circle the regular polygons.

SRB
104

---

6. Complete.

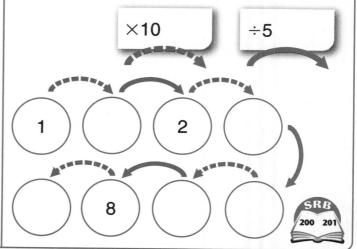

SRB
200 201

LESSON
6·8

Marking Angle Measures

Connect 2 straws with a twist-tie. Bend the twist-tie at the connection to form a vertex.

◆ Place the straws with the vertex on the center of the circle.

◆ Place both straws pointing to 0°.

Keep one straw pointing to 0°. Move the other straw to form angles.

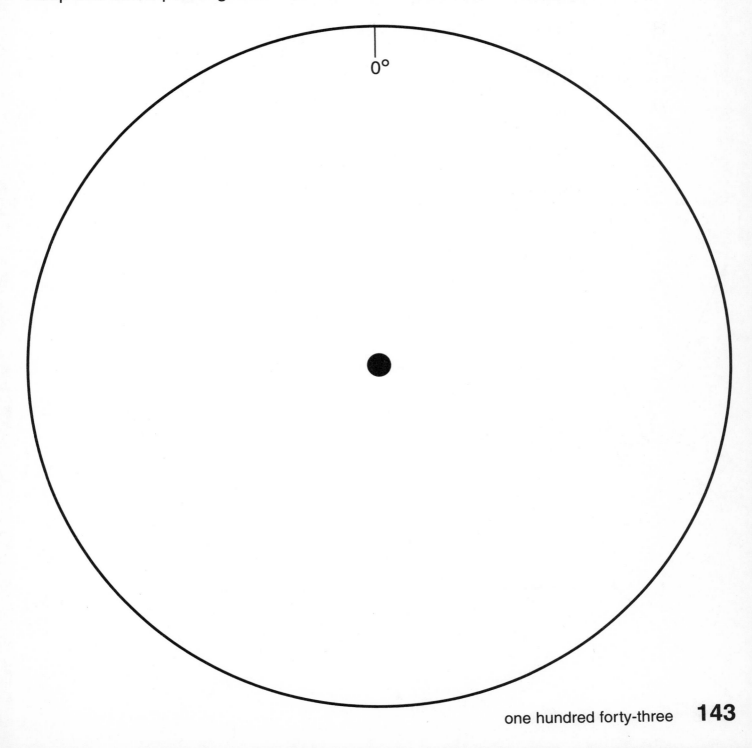

0°

LESSON 6·8   Measuring Angles

Use your angle measurer to measure the angles on this page.
Record your measurements in the table. Then circle the right angle below.

Angle	Measurement
A	about _____ °
B	about _____ °
C	between _____ ° and _____ °
D	about _____ °
E	about _____ °
F	about _____ °

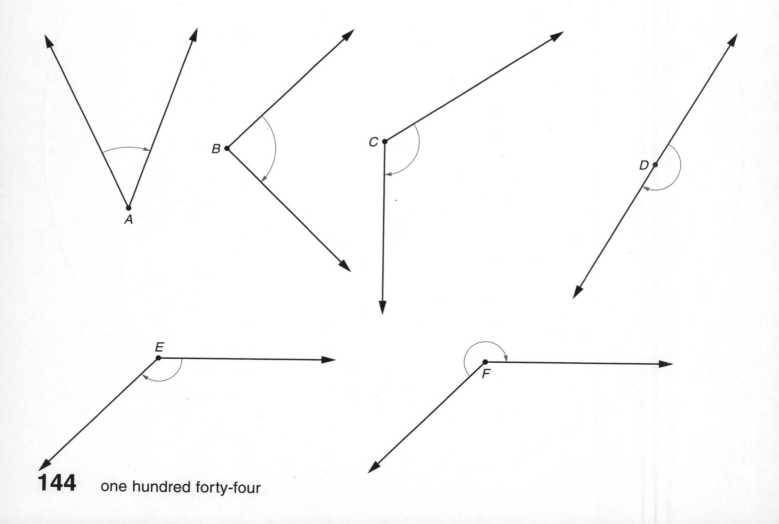

LESSON 6·8 **Math Boxes**

1. Continue the pattern.

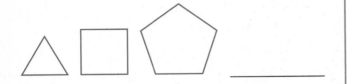

_____

SRB
197

2. If each grid is ONE, what part of each grid is shaded? Write the decimal.

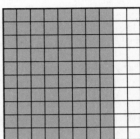

_____   _____

Circle the smaller number.

SRB
34 36

3. Write these numbers in order from smallest to largest:

0.2; 0.02; 0.19

_____   _____   _____

smallest   largest

SRB
36

4. Write a number model that matches the diagram.

vans	people per van	people in all
7	7	49

Number model: _____

SRB
259

5. Write the letter that names the right angle.

_____

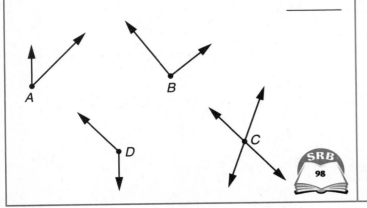

SRB
98

6. Complete the Fact Triangle. Write the fact family.

_____

_____

_____

_____

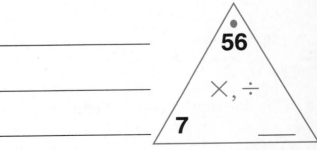

SRB
55

LESSON 6·9   Symmetric Shapes

Each picture shows one-half of a letter. The dashed line is the line of symmetry. Guess what the letter is. Then draw the other half of the letter.

1.   **2.**   **3.**    **4.**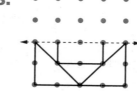

Draw the other half of each symmetric shape below.

5. 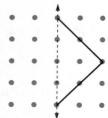   **6.**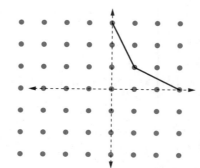

7. 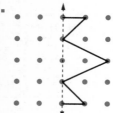   **8.**

9. The picture at the right shows one-fourth of a symmetric shape, and two lines of symmetry. Draw the mirror image for each line of symmetry.

Try This

10. The finished figure in Problem 9 has 2 more lines of symmetry. Draw them.

LESSON 6·9 **Math Boxes**

---

1. 3 people share 14 pennies.

Each person gets _____ pennies.

There are _____ pennies left.

73 74

2. A baker packed 8 boxes of cupcakes. She packed 4 chocolate and 4 white cupcakes in each box. How many cupcakes did she pack in all?

_____
(unit)

250–253

---

3. Draw a quadrangle with exactly one right angle. Label the vertices *A, B, C, D.* Which letter names the right angle?

Angle _____

98
108 109

4. Use your template. Draw a shape that has 6 vertices.

This shape is a _____

103

---

5. Describe the angle.

Fill in the circle for the best answer.

◯ A. greater than a $\frac{1}{4}$ turn

◯ B. less than a $\frac{1}{4}$ turn

◯ C. greater than a $\frac{1}{2}$ turn

◯ D. one full turn

168

6. Estimate. A package of cookies costs $2.09. About how much do 3 packages cost? Show the number model for your estimate.

About _____

Number model:

_____

SRB
191 193
194

LESSON 6·10   Base-10 Block Decimal Designs

Exploration C:

Materials   ☐ base-10 blocks (cubes, longs, and flats)
                   ☐ 10-by-10 grids (*Math Journal 1,* p. 149)
                   ☐ crayons or colored pencils

Think of the flat as a unit, or ONE. Remind yourself of the answers to the following questions:

◆ How many cubes would you need to cover the whole flat?

◆ How much of the flat is covered by 1 cube? By 1 long?

Follow these steps:

Step 1   Make a design by putting some cubes on a flat.

Step 2   Copy your design in color onto one of the grids on journal page 149.

Step 3   How much of the flat is covered by the cubes in your design? To help you find out, exchange as many cubes as you can for longs.

Step 4   Figure out which decimal tells how much of the flat is covered by cubes. Write the decimal under the grid that has your design on it.

Repeat steps 1–4 to create and count other designs.

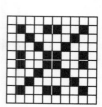

Step 1

Step 2

Example:

Step 1: Make a design on a flat with cubes.

Step 2: Copy the design onto a grid.

Step 3: Exchange cubes for longs. Figure out how much of the flat is covered.

Step 4: Write the decimal under the grid.

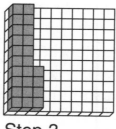

Step 3

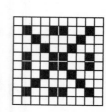

Step 4
Decimal: 0.24

LESSON 6·10

10 × 10 Grids

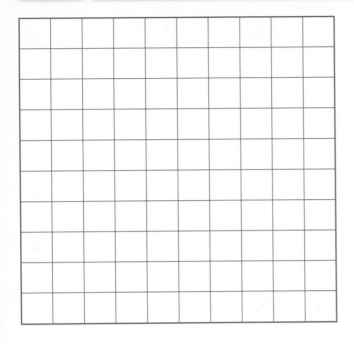

Decimal: _____

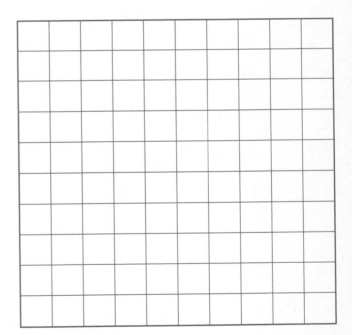

Decimal: _____

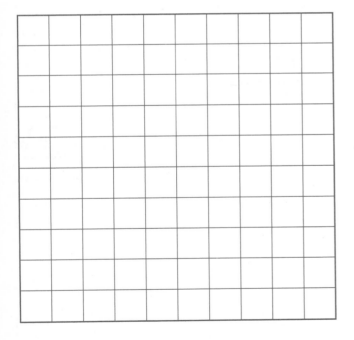

Decimal: _____

Decimal: _____

LESSON 6·10   Math Boxes

1. Draw a line segment, $\overline{DI}$, parallel to the line, $\overleftrightarrow{PO}$. Draw a *ray, $\overrightarrow{LA}$,* that intersects the line, $\overleftrightarrow{TW}$.

P ● ————————— ● O

T ● ————————— ● W

SRB 100 101

2. These letters are *Symmets:*

H, T, M, A

These letters are not *Symmets:*

F, J, R, S

Write other letters that are *Symmets:*

_____

_____

SRB 122 123

3. What is the difference in points between Players B and C?

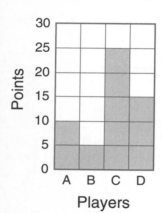

_____ points

What are the total points scored for all players?

_____ points

SRB 86 87

4. Write the numerals.

forty-hundredths _____

four-tenths _____

six-tenths _____

sixteen-hundredths

_____

SRB 33 34

5. Connect 4 points. Label the points.

What shape did you draw?

_____

SRB 108 109

6. Multiply.

$2 \times 5 =$ _____

$7 \times 3 =$ _____

_____ $= 5 \times 5$

_____ $= 2 \times 7$

_____ $= 4 \times 6$

SRB 52 53

LESSON 6·11 **Symmetry**

If a shape can be folded in half so that the two halves match exactly, the shape is **symmetric.** We also say that the shape has **symmetry.**

The fold line is called the **line of symmetry.** Some symmetric shapes have just one line of symmetry. Others have more.

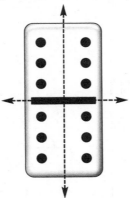
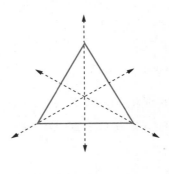

1 line of symmetry   **2 lines of symmetry**   **3 lines of symmetry**

1. Which of the following shapes is **not** symmetric? _____

    a.

    b.

    c.

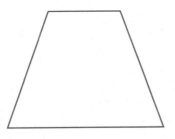
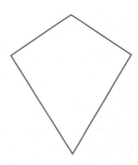

    d.

    e.

    f.

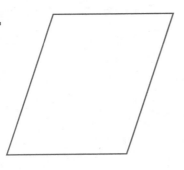

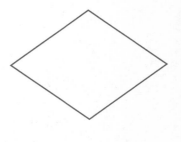

2. Draw all the lines of symmetry on the shapes that are symmetric.

LESSON 6·11 Math Boxes

1. 4 people share 18 crayons.

Each person gets _____ crayons.

There are _____ crayons left.

SRB
64 65
73 74

2. Dale had 9 toy cars. Jim had 4 less than twice as many as Dale. How many toy cars did Jim have?

_____
(unit)

3. Circle the right triangles. Use the corner of a piece of paper to check.

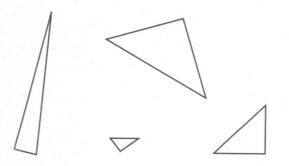

SRB
107

4. Solve the riddle.

I have four sides. My opposite sides are equal in length. One pair of my sides is longer than the other pair. Draw my shape.

I am a _____.

SRB
109

5. Fill in the circle for the best answer. The turn of the angle is

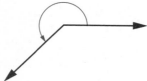

○ A. greater than a $\frac{3}{4}$ turn.

○ B. less than a $\frac{1}{4}$ turn.

○ C. greater than a $\frac{1}{2}$ turn.

○ D. a full turn.

SRB
168

6. Estimate. 1 bag of marbles costs $1.45. About how much do 2 bags cost? Show the number model for your estimate.

About _____

Number model:

_____

SRB
191 193
194

LESSON 6·12   **Pattern-Block Prisms**

Work in a group.

1. Each person chooses a different pattern-block shape.

2. Each person then stacks 3 or 4 of the shapes together to make a prism. Use small pieces of tape to hold the blocks together.

3. Below, carefully trace around each face of your prism. Then trace around each face of 2 or 3 more prisms on a separate sheet of paper. Ask someone in your group for help if you need it. Share prisms with other people in your group.

LESSON 6·12

Math Boxes

1. Draw a line, $\overleftrightarrow{AB}$, parallel to line segment, $\overline{CD}$. Draw a ray, $\overrightarrow{EF}$, that intersects the ray, $\overrightarrow{GH}$.

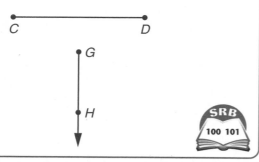

SRB
100 101

2. Draw all the lines of symmetry.

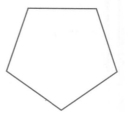

There are _____ lines of symmetry.

SRB
122 123

3. Number of days for one revolution around the sun:

Mercury	88
Venus	225
Earth	365
Mars	687

Which planet takes the fewest days to revolve around the sun?

Fill in the circle for the best answer.

○ A. Mercury   ○ C. Venus

○ B. Earth   ○ D. Mars

SRB
79

4. Write the numerals.

five-tenths _____

five-hundredths _____

three-tenths _____

three-hundredths _____

SRB
33 34

5. Connect 3 points to make a right triangle. Label the points.

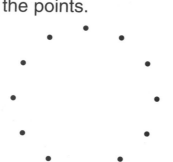

Which letter names the right angle? _____

SRB
106 107

6. Divide.

$30 \div 6 =$ _____

$12 \div 4 =$ _____

$20 \div 5 =$ _____

_____ $= 14 \div 7$

_____ $= 9 \div 3$

SRB
52 53

**LESSON
6·13**   **Math Boxes**

---

1. Solve.

$2 \times 2 =$ _____

$5 \times 5 =$ _____

$3 \times 3 =$ _____

$4 \times 4 =$ _____

2. Solve.

Double 3 _____

Double 30 _____

Double 300 _____

Double 7 _____

Double 70 _____

Double 700 _____

---

3. Solve.

$5 \times 4 =$ _____

$2 \times 7 =$ _____

_____ $= 3 \times 10$

_____ $= 7 \times 10$

$3 \times 5 =$ _____

4. Write 4 multiplication facts you need to practice.

_____

_____

_____

_____

---

5. Write 4 division facts you need to practice.

_____

_____

_____

_____

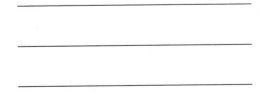

6. Complete the Fact Triangle. Write the fact family.

_____

_____

_____

_____

45

$\times, \div$

5 _____

Date _____ Time _____

Notes

Date _____ Time _____

Notes

Notes

Date _____ Time _____

Notes

LESSON 4·6 Multiplication/Division Fact Triangles 1

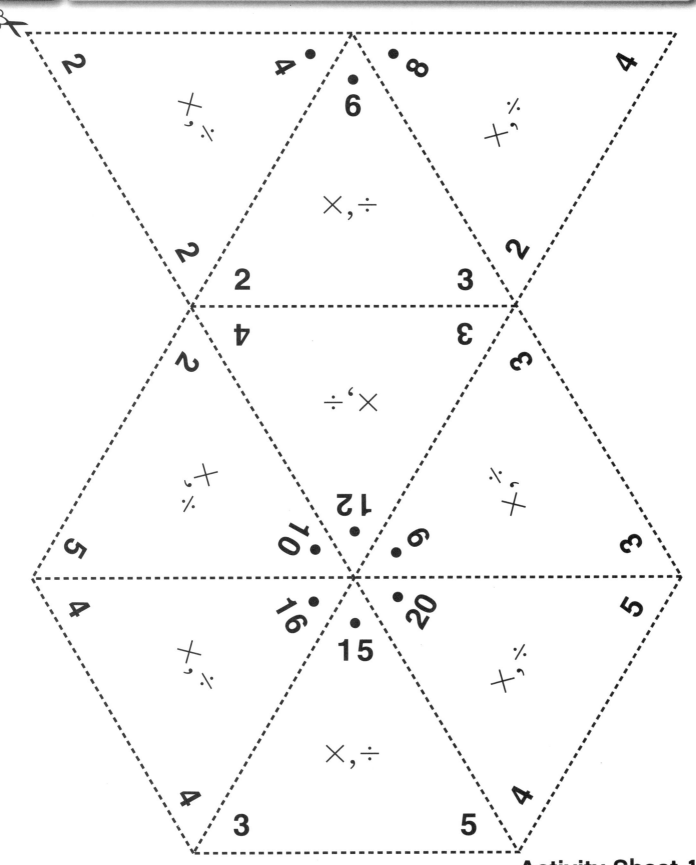

Activity Sheet 1

LESSON 4·6 Multiplication/Division Fact Triangles 2

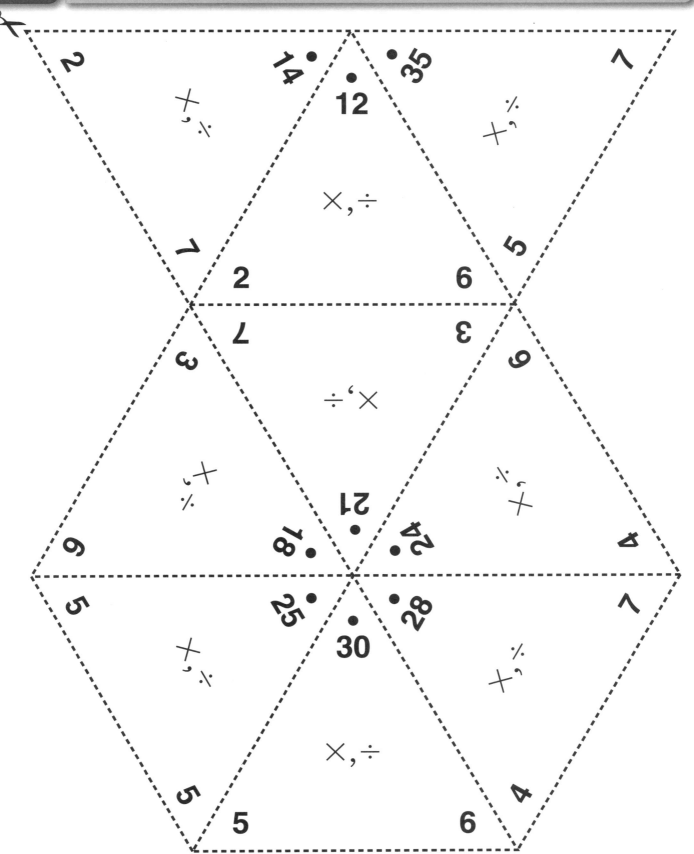

Activity Sheet 2

LESSON 7·2 ×, ÷ Fact Triangles 3

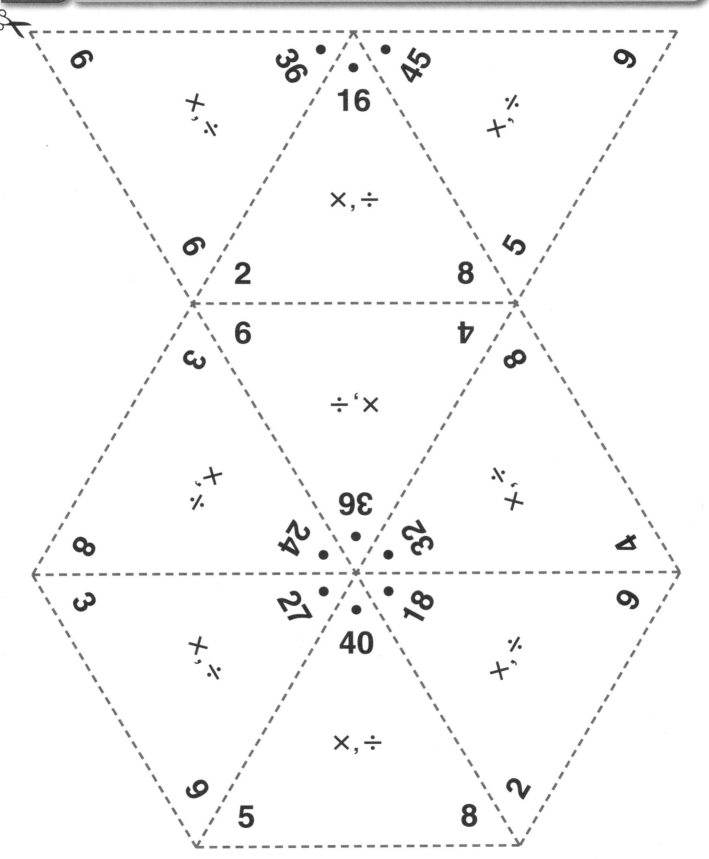

Activity Sheet 3

LESSON 7·2 ×, ÷ **Fact Triangles 3**

5 by 9

6 by 9

2 by 8

4 by 9

4 by 8

3 by 8

2 by 9

3 by 9

5 by 8

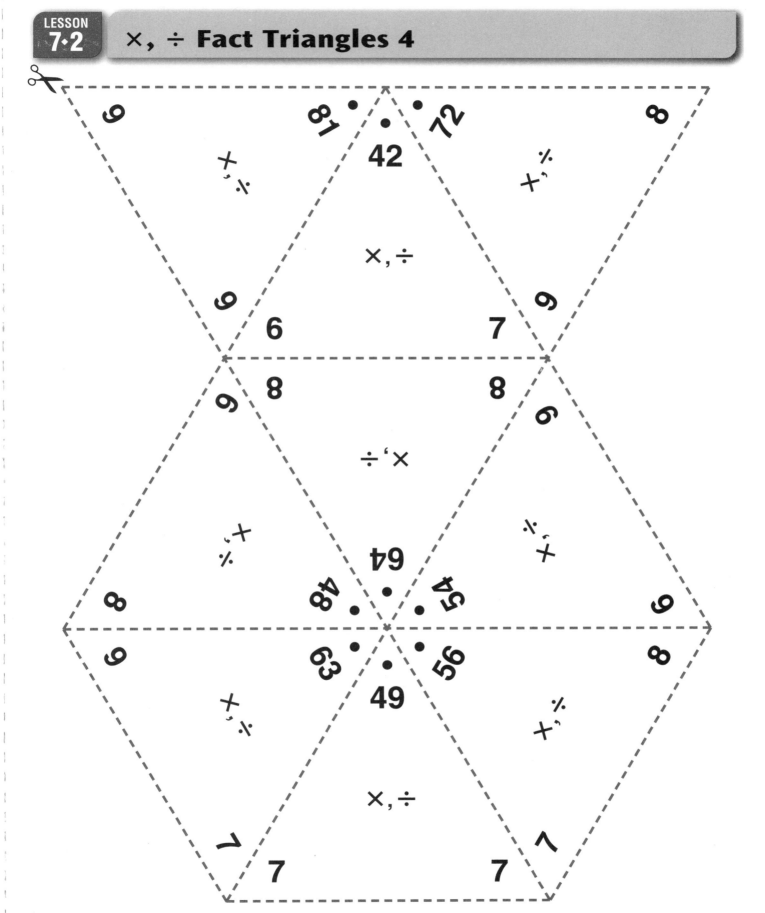

Activity Sheet 4

LESSON 7·2   ×, ÷ Fact Triangles 4

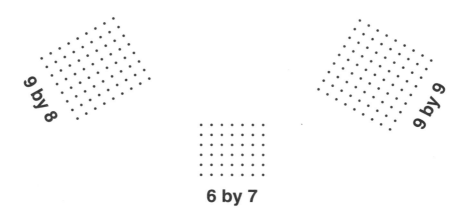

6 by 7

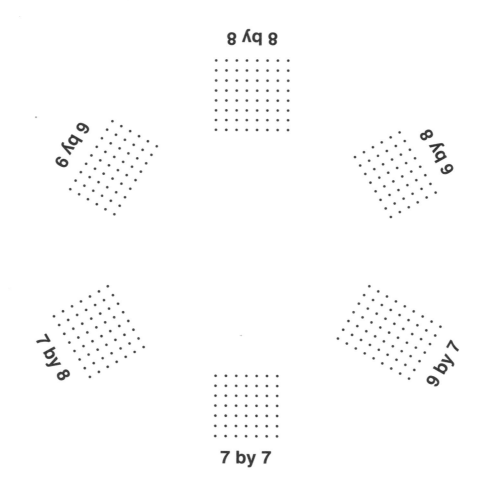

7 by 7